Logging

Ponderosa pine near Lava River Cave, Newberry National Volcanic Monument near Bend, Oregon.

—Photo by Bert Webber, 1991

Falling a ponderosa pine in Eastern Oregon by hand saw – date unknown.   —Photo from Bert Webber collection

This Is Logging

and Sawmilling

—DOCUMENTARY—

Includes:
Making A Logging Camp Work
Logging by Helicopter
Railroading in the Woods
Logging in the Snow
Fighting Forest Fires

Bert and Margie Webber

WEBB RESEARCH GROUP PUBLISHERS
Books About the Oregon Country

Published by:
WEBB RESEARCH GROUP PUBLISHERS
P. O. Box 314
Medford, Oregon 97501

Cover pictures:

(FRONT) Sugarpine log from Hoxie-Griffin BLM Sale near Ashland, Oregon.
Photographed February 1, 1996 by Bert Webber
(BACK) DC-7 air-tanker, U. S. Forest Service
fire-fighting mission in Deschutes County, Oregon June 8, 1992
Photograph courtesy of Medford Fire Center Archives

About the photographs:
Unless otherwise credited, the photographs were
made by the authors over a fifty year period
or are from the authors' collection.

Library of Congress Cataloging-in-Publication Data:
Webber, Bert
    This is logging and saw milling : documentary / Bert and
Margie Webber.
        p.   cm.
    Includes biographical references (p. 151) and index.
    ISBN 0-936738-18-9 - pbk
    1. Logging — Northwest, Pacific. 2. Forest fires — Northwest,
        Pacific — prevention and control. I. Webber, Bert. II. Title.
    SD538.2.N75W435 1996
    634.9'8–dc20   96-12172
                                                                         CIP

Additional subject headings for regional librarians:
Making a logging camp work / logging by helicopter / logging by
railroading / logging in the snow / fighting forest fires.

Table of Contents

Cedar logs at Rogge Forest Products Company log yard, Bandon, Oregon.   —Photo by Bert Webber *ca* 1975

Introduction

The timber business has traditionally been one of the top industries of the Pacific Northwest but how many logging companies and saw mills there have been in the region's history does not seem to be known. There does not seem to be an agency that keeps track of this but the total must be huge.

While making a recent aerial photo study of some lumber mills just in Josephine and Jackson Counties of Oregon, we photographed over 20 mills within 90 minutes flying time from Ashland to Grants Pass then south flying over highway 199 as far as Rough and Ready Lumber Company near Cave Junction. Due to technical limitations, we did not photograph anywhere near all of the mills that were below us.

In 1994, the latest year for which there are statistics, there were 166 saw mills in Oregon. This does not include mills doing other things with lumber originating from these saw mills such as veneer, plywood, particle board, door and window frames, woodchip makers, etc.

Over the years, these logging and mill operations have ranged from one-man entrepreneurial enterprises to huge entities that hired hundreds of men at a time.

The impact on the timber business in the Pacific Northwest, as elsewhere, by the environmentalists, has been devastating. Anyone who doesn't want to think so, merely needs to look at the number of mills that now operate part time, or have closed entirely, in recent years due to shortage of logs. The partial or complete shutdowns have thrown thousands out of work and seriously disrupted local economies. Some of the biggest, as Weyerhaeuser, closed their huge mill at Bly, Oregon and removed all the buildings. The firm also operated a railroad between Klamath Falls and Bly and now the railroad, track and all, is gone.

In Medford, MEDCO (the Medford Corporation), long a pillar of the community noted for hiring thousands over the life of the firm, is today only a vacant lot where the mill once stood. And its railroad, track and all, is gone.

The One-Log Load

One-log loads might be seen today if one is lucky but, for the most part, the days for viewing such logs are about gone.

The log on the flatbed truck was prepared by the State of Oregon Department of Forestry for exhibit in the Jackson County Fair in 1987.

The sign reads:

Old Growth

There are more than 742,000 acres of protected old growth forest in Southern Oregon. It will never be cut.

This Douglas fir log, harvested from commercial forest land on the Tiller District of the Umpqua National Forest, is 675 years old.

Its vital statistics are:

Weight: 576, 860 pounds

Length: 38 feet

Diameter: 87-inches (7¼ feet) large end
70-inches (approx 5¾) small end

Net Volume: 8,350 board feet

Lumber Volume: Enough to construct 1.17 average-size homes

Plywood Volume: Enough ⅜ inch plywood to meet the requirements of 5.39 homes

Veneer Volume: Enough 1/10 inch veneer to make one sheet 8-feet wide and 2.63 miles long.

There are mills that continue to thrive but at various degrees. Some operate whenever they can get logs while others stockpile acres and acres of logs in their mill yards. A number of these cold decks can be seen from nearby highways. Others are best viewed on a flyover at low altitude. We have looked at both full and empty log yards from the ground and from the air.

Many books published in the past that deal with logging seem mostly to be filled with the oldest days of the industry – oxen or horse teams pulling (yarding) logs from the falling sites to the landings, then the use of steam-powered donkey engines doing likewise. Great books! But the industry has changed. Accordingly, this book opens immediately with the latest logging technique: helicopter logging. Further, in years past, most logging outfits shut down with the approach of the winter snows. Here we present another recent concept in our chapter "Logging in the Snow."

To learn about helicopter logging, we sought guidance from Erickson Air-crane and from Croman Corporation. Each willingly provided information and pictures for which we thank them.

Few books on logging include much if anything about the greatest peril to the industry (environmentalism excepted) – forest fires. Our chapter "Fighting Forest Fires" brings to light the role of the U. S. Forest Service and the flying air-tankers that "bomb" forest fires with retardant. We are indebted to Phil Cardin, U.S.F.S. Forest Dispatch Coordinator at the Medford Fire Center, for inviting us to view his system and see how it worked.

On the subject of air-tankers, we were around a number of years ago when a B-24 landed in a mountain lake. The various reporters presented their versions at the time, but over the years the stories grow. It was our good fortune, while this book was in progress, to meet Gary Austin who was the co-pilot. Gary told us what really happened and then loaned a number of unique photographs of the B-24 which was technically a PB4Y2. Some of them are included here. We thank Gary for his timely availability and for his interest and aid.

We found a saw mill whose management welcomed our request to view and photograph the operation from logs to finished lumber. We thank General Manager Ryle Stemple of the Central Point Lumber Company for his interest, guided tours and enthusiastic assistance.

Of great help was Edsil Hodge who came out of retirement to tell us of some of his 50-year experiences working in the specialty mill at the House of Myrtlewood in Coos Bay. This unique myrtlewood mill is unknown to most people but we are able to include something about it here.

Our long-time friend John T. Labbe, a published author many times over, who has particular

Collier Memorial State Park

On Highway 97 in Klamath County about five miles north of Chiloquin

This park is the major logging museum in Oregon. It is in a ponderosa pine forest at the junction of Spring Creek and Williamson River.

There is a full-service camp ground for overnighters in summer months. Fishing access.

The major attraction is the logging museum where antique equipment is on display. (In winter, these may be covered by snow.) The exhibits include a famous McGiffert loader, early log trucks, steam engines, tractors, etc.

The facility is maintained by the Oregon Department of Parks and Recreation.

knowledge about geared locomotives and donkey engines, provided some pictures we have included. We thank John for his willing assistance.

We appreciate those members of the Central Point Fire Department, and the Fire District No. 3 Jackson County Fire Department, who were available to talk with us about the time when a huge log deck caught fire in the middle of their town. We very much appreciate their willingness to loan photographs of the fire. Orville Eary, Protection Supervisor Medford Unit of the Oregon Department of Forestry, assisted with more details.

It is important to acknowledge the great professional assistance received from the corps of Reference Librarians at the Medford Branch of the Jackson County Library System. Who else can help locate or verify finely pointed details but a professional librarian? Dr. Anne Billeter (Ph.D.) heads this department. We sincerely thank them.

The majority of the content of this book starts about 1920 and runs right up to the present time. It would be impossible to be serious about putting something about every woods operation and every mill in the Pacific Northwest into a single book. Such a book would take years to compile, could never be complete and would be so big it would be unaffordable by the public. It is our hope that the several timber operations and mills we have included, along with the many photographs, will be found representative of a tremendous industry, and will be acceptable to our readers.

Readers may send constructive criticism to us in care of the publisher whose address is found on page *iv*.

Bert and Margie Webber
Central Point, Oregon

Common Woods Terms

A Selected List

Cold deck	Iron ox	Kiln	Skid Road
Donkey	Locomotives	Kitten	Spar tree
Gang saw	Climax	Landing hooker	Ukulele
Hibernate	Heisler	Lay out	Vegetable fireworks
High lead	Lima Shay	Limb wood	Wigwam burner
Iodine	Willamette Shay	Lumberjack	Wobblies
			Yarding

(LEFT) **Cold deck of logs cut to length for a veneer peeler lathe.**
(RIGHT) **Veneer awaiting transport to plywood lay-up plant.**

Cold deck

Logs piled in a yard in an orderly manner to make access to them, for later moving the logs to the mill, easy. Logs stacked in an area away from the woods, as in the yard of a sawmill. In early days, a large pile of logs dumped at most any angle sometimes forming a pyramid around a swing tree, or spar tree, in the woods.

Donkey

"Donkey" was the original name given to a small steam engine that a former seaman, Dolbeer, had adapted to work a capstan taken from a ship. With a line made from manila rope, he hauled logs in a timber operation near Eureka, California about 1888. This contraption was an early day winch. The manila rope eventually gave way to the much stronger steel cable, also called "wire rope." Early "Dolbeers" had just a short line for pulling logs from the forest to a skidroad where animals took over. Eventually loggers used long wire cables and with the steam-powered "donkey," which replaced four-legged donkey power, hauled logs all the way from the woods to the landing. The first donkey engine believed to have been used in the Pacific Northwest was on an operation on the Wishka River in Grays Harbor County of Washington in 1889.

9

The logging industry, in the early days, was manpower intensive helped with teams of oxen for yarding the logs. The industry was revolutionized when the donkey engine came along.

Donkey engine without smoke stack.

There were many varieties of donkey engines. Early models used wood cut from the forest as fuel to generate steam. But gasoline, diesel oil and electric powered donkeys came along. Not only did this heavy-duty which haul logs, it could move from position of use to a new location. By mounting the donkey on skids, and running the line to a sturdy tree, taking up the line on the drums with the winch, that had replaced the single capstan, the donkey was pulled through the woods to its new location.

Donkey engines grew to great proportions with the continuing need for more power. It has been reported that as many as 26 different specifications of steam powered donkey engines were built by a single manufacturer in the great northwest. There were many brands. These included: Portland, Seattle, Tacoma, Vulcan, Washington Iron Works, Willamette Iron and Steel Works and Skagit Steel and Iron Works. So popular were the donkeys, that it is reported that Willamette in Portland filled 51 factory orders in 49 days in 1913 alone.

It should not be assumed that the donkey engine was restricted to hauling logs from the forests. Steam donkeys were also found on ship decks for moving pallets of freight from the docks to shipboard and at the next port, for the unloading chores. During the Spanish-American War, while the battleship *Oregon** was on its speed run from San Francisco to Cuba to join the fight, valuable time was lost when hundreds of laborers had to be hired, on two occasions, to form a procession for hauling bags of about 50 pounds each of coal from a barge up over the side of the ship. If there had been a steam donkey on the deck of the *Oregon*, the re-coaling could have been done in just a few hours.

* For the entire amazing story of the battleship *Oregon*, refer to bibliography for *Battleship Oregon; Bulldog of the Navy.*

Gang-saw

One man could cut a lot of lumber with a gang-saw. Logs headed for this saw usually had their opposing round sides cut off to provide flat surface for easy transit on the saw's bench. The gang-saw was a very noisy operation with as many as 16 blades moving rapidly up and down "chonk-uh-chonk-uh-chonk" at one time.

Long before salvage of sawdust for resale came about, sawdust accumulated very fast around a gang-saw.

Hibernate

Loggers, most of them, who left the woods to shack up in town for the winter. Literally, to live in a certain place. (*slang*)- for opposite sexes to live together without the benefit of matrimony. It was a recognized practice for many women who worked in the camps in summer to seek a mate for the winter. He provided the facilities and board; she to provide all "wifely" duties.

High Lead (*See also* Spar tree)

A tall tree (spar) is selected near a landing. The highlead handles the main line and haulback cables high above the ground. The idea came by observing the rigging on sailing ships. With the cables high above the ground, it replaced the ground or low lead technique where the logs were dragged on the ground with all sorts of risks, the more common being when the pulled log became hung up on obstructions – rocks and stumps. With the high lead, the logs traveled through the unobstructed air. A side value in moving the logs through the air, the risk of hitting hidden yellow jacket nests and the continuous generation of clouds of dust the dragged logs created were eliminated.

Iodine

Strong coffee

Spar tree at East Side Logging Company in 1921.

Iron Ox

A geared locomotive without any cars and going at super-slow speed, when pulling an extremely large log behind it with a cable. The log bumped along between the rails to the landing.

Kiln

A kiln (pron: *kil*) is literally a stove. In the timber business, green lumber was "cooked – dried out" in heated fairly tightly constructed buildings. Some kilns had electric heating coils, others used natural gas burners. There were many fans to circulate the hot air through the stacks of lumber. The lumber was treated up to 48 hours in the kiln before being delivered to a customer.

Kitten

A very small tractor as opposed to a large tractor (cat).

Landing hooker

The man in charge of loading logs on to whatever conveyance was being loaded. The foreman

Lay out

A driver of a log truck, the load suspected of being overweight, who hides his load plausibly on a side road awaiting the evening closing of the state highway department scale house.

Limb wood

Branches from a tree cut to firewood length. Limbwood is often more dense than wood from trunks. Limbwood makes a hotter fire. Chief cooks liked it for use in the kitchen stoves.

Lumberjack

A loose word meaning a logger. The term was used by fiction writers who should have just said "logger" if they were writing about a man working in the woods.

Even with the great care used in choosing a spar tree, now and then, due to the sudden stresses placed upon it, the top would snap off. When this happened, it was a serious matter to re-rig lower on the spar or find a new tree close by because operations upon which the high line depended, were shut down until a "fix" was made.

Heisler

This Heisler locomotive has been reconditioned and operates on the Sumpter Valley Railroad in Baker County Oregon during summer months when it hauls a sight-seeing train from Ewan to Sumpter Valley Dredge State Park at Sumpter.

Locomotives (Major types)

The major difference between *road* locomotives and *logging* locomotives is that the commonly known road loco was rod driven. This means that the power from the cylinders was applied only to the drive wheels under the locomotive. Logging locomotives were gear-driven with power to every axle including those under the tender. The short wheel base of geared locomotives brought all the available power to *all* of the wheels and was especially appreciated on steep and sometimes rickety-laid temporary track in the mountains. The geared locos, with their short wheel base, were very flexible for they were designed to stay on the rails on very sharp curves which rod engines could not negotiate. The geared engines were noted for their superb pulling power at slow speeds. There were a number of manufacturers of geared locomotives but the most popular were - Climax - Heisler - Lima Shay - Willamette (Pacific Coast) Shay.

Climax

Climax:

The Climax Locomotive Works of Pennsylvania started production of a geared locomotive in 1888. The cylinders were parallel to the frame of the locomotive inclined at an angle of 45 degrees. The drive was similar to a ship's propeller shaft to the geared axles. This was a high-powered locomotive that was a favorite on many timber operations where the track was steep and had sharp turns.

Heisler

Charles Heisler's geared locomotive proved to be a challenger for quality right alongside that of Climax and Shay. His system included the features and power of the other locomotives but his was the faster of the three on straight track usually reserved for road (rod) engines. The Heisler had its cylinders mounted at 45 degrees set at right-angles to the frame. Power to the wheels was by connecting rods to the geared axles. There were side rods to all axles under the engine and tender.

Shay – Lima Shay and Willamette Shay

Each of the geared locomotives had a unique appearance. The Shay seemed to be lopsided when viewed from the front as the boiler was off-center. The weight on the trucks was equalized by the cylinders which were on the opposite side (the engineer's side). A heavy shaft, driven by cranks, was also on the engineer's side of the locomotive. It was geared to each axle on the engine as well as on the tender. The Shay became the most popular of the geared locomotives. It was designed by Ephraim Shay, a logger from Michigan, in 1874. Manufacturing was by the Lima Locomotive Works in Ohio, which built almost 3,000 of them as recently as 1944. Lima Shay locomotives could be ordered from a catalog offering different weights and values of power and other specifications.

Willamette Shay   (Pacific Coast Shay)

After the patent on the Lima Shay ran out in 1898, several imitation models appeared from several manufacturers. Of these, the most worthy was Willamette Iron and Steel Company in Portland. Officials there were well aware of the success of the Lima Shay among timber operators in the Pacific Northwest therefore, they reasoned it would be profitable to build their version of the Lima along with other loggers' products, as the donkey, in Portland. Willamette specifications followed those of the Lima Shay however Willamette's engineers made selective modifications. Among these were headlights and, in tribute to the northwest's rainy weather, a closed cab. In the case of East Side Logging Company, which developed its own specs, Willamette built to East Side's design. Some timber operators in the northwest sought the Willamette over the Lima at least partially because of the closer availability of replacement parts. In addition, when they had a bad day and wrecked a locomotive, the factory was nearby for complete rebuild. The Willamette Iron and Steel Company built 33 geared locomotives between 1922 and 1929. As of March 1, 1996, there are six Willamette Pacific Coast Shays extant. Five are publically exhibited in Bonner, Mont.; Cathlamet, Wash.; Port Angeles, Wash.; Medford, Ore.; Dunsmuir, Calif. One is privately owned by Jim Gertz near Port Angeles.

Skidroad

A right-of-way over which logs were "skidded," that is, pulled by animals (later donkey engines) on small sapling skids laid crossways on the roadway. It did not take long to abandon the skids thus the logs were merely pulled along in the dirt. In snow-logging, the skidroad is on a cushion of snow which protects the forest floor.(See chapter on "Logging in the Snow.")

The original skidroad was in Seattle where Henry Yesler laid a skid road between his mill on the waterfront, into a stand of timber at the top of a hill. This was in 1852. When Yesler no longer needed his road for hauling logs, saloons and stores, flop houses, hiring halls, card room and houses of ill repute sprang up along it. The neighborhood was known for its rowdiness where loggers hung out. A corruption of "skidroad" is the non-word "skidrow" of which there is no such thing or place, but seems to have originated with uninformed newspaper reporters. Yesler's original skidroad has become Yesler Way in downtown Seattle.

Spar tree

Rigging is hung for any of many cable hauling systems in the woods or at a mill. The spar tree is usually in the center of operations. In the woods, a tall, straight, live tree is chosen but in areas without such trees, a log, from elsewhere was brought to the site and erected. If the spar tree was live, it might also be called a "standing tree."

Erecting a spar tree at the Goodyear-Nelson Hardwood Company Mill No. 2 in Sedro-Woolley, Washington in 1946. In early days, a donkey engine would do the work. For the job shown, the spar was raised with the winch on a D-8 Caterpillar tractor.

Ukulele

Shovel with a short handle.

Vegetable fireworks

Beans which create flatulence

Wigwam burner – Consumer

The waste from a saw mill – odd pieces of lumber, trim ends and sawdust – had to be disposed of. In the early days, fires in huge pits did this job but the risk of an open fire getting away and burning down the forest led to the invention of the wigwam burner, also called a "consumer." These burners were usually made of sheet metal. The waste materials to be burned (consumed) were conveyed into the burner on a conveyor. The fire was generally started by dousing a load of waste with oil. Sometimes, during light mill production periods, to keep the fire going, old automobile tires were burned. The fire in a consumer generally only smoldered with the

continual feeding of mill waste. During World War-II, when blackouts were ordered, passing aircraft could still determine the locations of the mills by the dull red glow from the wigwams. In addition to discontinued feeding of the consumers at night, a new system of automatic sprinkling of water on the fire stopped the glow, but this dousing was not enough to put out the fire. In the morning, the fire was easily brought back to full strength.

The object of most manufacturers, whether in the timber business or otherwise, is to get full usage out of raw material. Where there was traditionally much waste after felling a tree, today there is very little. The scrap trim wood and saw dust is presently marketed in one way or another. Much sawdust eventually went into Presto-Logs for fireplaces, and sold to makers of charcoal briquettes. Scrap wood becomes particle board, or may be chipped with much of it going to paper making plants. Use of the wigwam burner is long gone and very few remain to be seen.

Wobblies

Anyone who was a member of the Industrial Workers of the World (IWW). This radical labor group, which functioned during and after the First World War, caused a measurable amount of distress to timber company management.

Yarding

In the earliest days of west coast logging, teams dragged logs over skids (hence skidding) from the forest to the log yard, also called the landing or dump. Horses or mules, sometimes oxen, were used to pull logs over the skids to the log yards – yarding. This was also called skidding or wheeling when the big wheel conveyances were employed to lift the lead-end off the ground so the logs would not dig in to the earth while being pulled. Though called hauling, this was a kind of yarding. Then Dolbeer developed the vertical spool steam donkey (DOLBEER DONKEY – photo page 71) which could drag logs with a cable from the forest to the landing. A horse, a mule or laborers returned the line back into the woods. About this time skidding became known as "yarding," or ground lead yarding, because the logs were hauled to the "yard" (landing). The addition of the haulback cable did away with the line animals. The highlead replaced the ground lead all of which were initially operated with donkey engines. Eventually, gasoline and diesel engines, even some with electric motors, replaced the steam donkeys. Later, "cats" (tractors) pulling arches replaced the draft animals and big wheels. Currently, "cats," using just a cable and a claw clamped to the log (see page 31), pull logs out of the woods on a skidroad to the landing. Regardless of the methods employed for hauling logs from the position where the tree was felled to the landing, the process is still known as yarding.

Tractor-skidding with Athey Wheel Cargo Logging Arch *ca* **1954.** (RIGHT) **Tractor skidding with D-6 cat February 1996.**

Helicopter Logging

From the first moment when you look up into the sky because of the odd sound coming from that direction, and see a helicopter of extra-ordinary size you might wonder what it's doing? Then a moment later you observe the unbelievable – long heavy logs sailing through the air dangling under the helicopter on the end of a long cable.

Of course you are viewing this scene away from town, as dangling logs from helicopters over the heads of people doesn't happen — Safety.

For decades mankind has been harvesting timber from the forests. As with any industry, the environment has begun to suffer the impact of this long term intrusion. It is this knowledge that takes a team of highly skilled professionals from Erickson Air-crane Company into the future – the future of successfully harvesting timber while not damaging the health and beauty of the forests.

From the first time you see one, it is clear that the S-64 Erickson Air-crane is no ordinary helicopter. This is the only helicopter ever specifically designed to perform precision *external* operations.

In 1971, when Jack Erickson was looking for a more environmentally sound method for harvesting timber, he studied the S-64 which had been originally designed by Sikorsky Aircraft.

With acquisition of the aircraft type certificate in 1992, Erickson Air-crane Company became the legal manufacturer of the S-64 Sky-Crane that had been modified and improved to become the Erickson Air-crane. Because of the great lifting capacity – 25,000 pounds – logging by helicopter has become economically feasible. Now, logging operations can be conducted from the sky instead of on roads hacked through the forests. Sky logging substantially lessens the impact of forest damage.

Helicopter logging can enable man to selectively remove timber with negligible shock to the surrounding area. Many different logging techni-

19

ques have been used in the past and some have had left a devastating impact to the forest and on the ground. However, selective logging by experienced forest crews, combined with aerial removal, allow timber to be lifted up and out of the forest with minimal damage to the surrounding area.

Natural disasters such as flooding, drought, hurricane, bug kill and forest fires claim millions of board feet of timber every year. Unless this timber is removed from the forest, it will dry out and can become a threat (bugs and fire) to the surrounding environment. While selective logging is an obvious answer to forest management, access and logistics of removing timber from sensitive and very remote areas has been an ongoing challenge.

Jack Erickson, a second generation logger and owner of the company that bears his name, realized the potential and need for aerial logging. He initiated the concept and pioneered the industry. Over 65,000 flight hours, and more than two decades of experience in the helicopter timber harvesting industry, have created the skill and expertise found in the company today.

The Erickson S64E Air-crane is the only heavy lift helicopter that can effectively fly uphill with its certified hook weight enabling it to always pick up timber and lift it out of the forest without dragging the logs through the fragile surroundings. The power is in the two 4,500 horsepower each jet engines which allows the helicopter to always operate with surplus power. It has immense power, speed and maneuverability and is guided by experienced pilots and ground support.

Landings for logs are one of the first consid-

(LEFT) **Logs, attached to cable, being lifted from the ground by helicopter. When clear, the helicopter** (ABOVE) **whisks them away to the landing.**

copter flies woods crew members within a close proximity to the work area thus eliminating the

(LEFT-TOP) **Support helicopter.** (LOWER) **Landing in the forest. Some timber companies negotiate with property owners to re-use an old landing that had been cleared years earlier.**

need for roads. Once cutting is completed on a given site, choker setters and strip runners take over. Chokers are gathered on the landing, coiled and separated into groups of ten to twenty, or less when needed, for special drops. The support helicopter transports the coiled chokers to the drop site. The drops are coordinated by radio between the choker setters on the ground, the pilot and a landing crew member where drop sites were laid out with fluorescent ribbon for easy identification. At times, a signal to the helicopter is made with a mirror signal for a precise location.

In some areas where removing smaller timber is the project, more choker drops are needed. The Erickson carousel hook enables as many as four separate choker drops at one time allowing exact placement of multiple drops. This way the choker setters can move quickly through the forest without having to carry heavy chokers by hand.

Chokers are secured around the logs and connected together to build a "turn" which will utilize the maximum life capabilities of the Erickson Aircrane. The choker setters are extremely experienced in weight calculation of timber to insure the best performance of the helicopter. Once enough turns are set, the wood crew members stand clear of the designated flight path, the traffic patterns having been carefully set out.

When all is ready, a hook tender takes his position at the first turn and the Air-crane begins the carefully laid plan of flying the timber out of the forest. The pilot and co-pilot are in constant radio communication with the ground crew. The pilot flies the hook to the hook tender. The hook is usually attached to the aircraft by a 200 foot steel cable called a long line.

The hook tender radios the pilot of estimated weight, number of logs and choker configuration. With great skill and precision, the pilot places the hook near the hook tender's hands then the tender quickly lays the choker nubbins in the hook and

erations to operating a cost effective helicopter program. Because the S-64 can effectively fly up or downhill with its loads, landing can be conveniently located in whatever direction is selected. Experienced timber fallers fall the trees in a designated direction to minimize damage to undergrowth and to make yarding easier.

After a tree is down, the cutters prepare it for the flight out. Part of this preparation is to cut the log into the sawmill's preferred lengths as well as measure the logs to determine the approximate board feet.

Access and punctuality go hand in hand. Access is never any trouble as heliports are built in units where steep terrain exists. A support heli-

The "Back-seat" Driver

The Erickson Air-Crane is the only helicopter with an aft (rear) facing pilot station. Looking out the rear of the cockpit, the aft pilot has an unobstructed view of the rigging and load. The front seat pilot flies the helicopter to and from the job but the back seat pilot, with full set of controls, lifts and sets each haul. The anti-rotation rigging is used to prevent the load from spinning which eliminates the need for tag lines that can pose danger to the ground crew.

leaves the area. Once the tender is "clear," as signaled by radio, the Air-crane picks up the turn and flies it out of the forest.

While the men on the ground estimated the weight of the haul, a load cell on the helicopter indicates the exact weight being lifted. The co-pilot advises the pilot of weights and instrument readings which are critical. Once the logs have cleared surrounding obstacles, the crane begins its speedy ascent or descent and heads for the log landing. Even with maximum weight, the crane can quickly reach speeds of more than 120 miles per hour for the short run to the landing.

Compared with tractor time moving through the trees along a skid road (about 5 miles per hour) with an equally sized load of logs, the heli-

from the long line. Within seconds, the helicopter is on its way back into the forest for another turn. On the ground, the chokers are taken off the landed logs and recoiled for the next need. Experts remove additional limbs from the logs and heavy equipment operators sort and stack the logs into decks. Then a loader operator places the logs onto 18-wheel log trucks for the final trip to the mill.

On the ground, radio communication is maintained between the drivers and loader operators on CB. These crews process as many as sixty truckloads of logs in a day.

A part of the Erickson operation is to assure prompt helicopter technical support by having a service landing site centrally located which is the hub of operations. Aircraft maintenance teams, with parts, are on location where aircraft mechanics perform whatever work is required on a moment's notice which minimizes down time and loss of production.

Logging by helicopter achieves a new standard of excellence in timber harvesting. Helicopter logging also preserves the environment.

Helicopter logging is here to stay. ◇

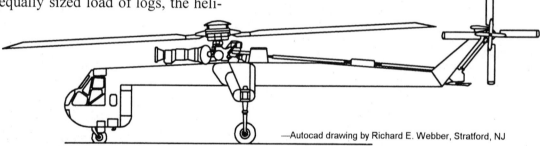

—Autocad drawing by Richard E. Webber, Stratford, NJ

copter method has proven very cost effective.

All members of the team work with effective speed and organization. At the landing, the logs are carefully lowered to the ground and detached

SIKORSKY S-64F SKY CRANE

Total number of aircraft produced: 90
Production years: 1966-1975
Engines: (2) Pratt & Whitney JFTD 12
Horsepower: 4,500 each engine
Weight: 22,000 pounds
Max. payload: 25,000 pounds
Crew 2   Passengers: 0
Fuel capacity 1,350 gallons
Cruise speed: 135 mph
Range: 350 miles
Unique feature: Rear (aft) facing pilot station for 3rd pilot when needed.

—Data from Erickson Aircrane, Central Point, Oregon

Logging in the Snow

Decades ago, most timber operators declined to work the forests in winter months due to the added costs caused by the often freezing weather. Further. the slop of mud and snow was risky to the men and could damage equipment.

As many operations depended on steam locomotives to pull the logs from the woods, maintaining the engines in the usual limited shop facilities, in the sub-freezing weather, was a near-impossible task, let alone the risk of derailments because of ice forming in the switches. And finding woods crews who would work in frigid weather proved to be a real chore as hiring halls were largely vacant in the winter time, the men having hibernated in the sunny south until spring.

Times have changed. In the dead of winter (January 24, 1996) with snow on the ground, timber fallers working for Croman Corporation started to cut on the Hoxie-Griffin Timber Sale (Bureau of Land Management). This is sixteen miles east of Ashland, Oregon. Croman went to work on the 2.8 million board foot sale on a 255-acre site.

In order to minimize environmental impact on the forest floor, Ken Brown, a BLM contract administrator said:

When you have enough snow [on the ground], the equipment can run across the snow without compacting the ground in terms of long-term damage. [When] the snow melts, the ground isn't harmed and you've protected the new stand that's coming up.

The terms of the contract state there would be no logging if there was less that 18-inches of snow as a protection to the young growth and the ground. When the cut started, there was close to four feet of snow on the ground with more snow in the forecast. The site is in the Cascade Mountains east of Howard Prairie Lake at about the 4,700 feet elevation.

In the face of environmental activists whose object is to stop logging everywhere, the Bureau of Land Management announced that guided tours would be provided for protesters and media repre-

This cradle in the snow was caused after a large sugarpine was felled and landed here. The stump is in the background at the end of the cradle.

(TOP) **Catskinner carefully works his 'dozer to logs in the forest that await skidding to the log deck at the landing.** (LOWER) **Because of the deep snow, the tread on the cat does not damage the forest floor, merely leaves "foot-prints" in the snow.**

Cleanest, fastest and safest cuts are made with sharp saws. Timber faller installs new chain when earlier chain has been dulled by accident (as when hitting a rock), or dulled by continuous use. When damage to a chain seems minimal, faller may sharpen (file) his own chains. Most timber fallers keep two spare chains in their gear when in the woods. Stihl saw shown has 36-inch bar.

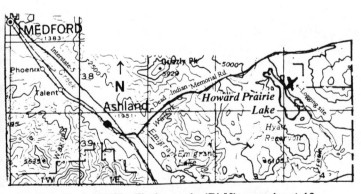

The Hoxie-Griffin Timber sale (BLM) was about 16 miles east of Ashland in Jackson County Oregon.

sentatives. A BLM spokesman stated, "We want this to be open, to let people see for themselves what is going on – we're willing to let them see any thing they want to see."

Because of the need for safety, the BLM said its foresters would guide people who are interested in viewing the work a few at a time weather permitting. "We don't want anyone hurt." During the first few days of the work, the temperature was varying between 25 and 29 degrees F.

The Jackson County Sheriff's Department put deputies at the access road passing through only individuals who had a permit to visit or who were accompanied by BLM or contractor personnel. The reasons for this control was clearly spelled out: S-A-F-E-T-Y.

Steve Armitage, forester for the BLM Ashland Resource Area, pointed out that the access road was narrow because of highly plowed snow banks on each side that caused some curves to have limited vision for on-coming traffic – loaded log trucks. "We'll take people in there as long as they stay with us and don't create a safety problem or hazard to the operation."

On the first day of the cut, about two dozen protesters showed up but they saw little. Instead of what some envisioned viewing, the wholesale sawing down of dozens of trees – clear cutting – and bulldozers smashing their way through the underbrush, this was a relatively quiet and carefully planned operation.

A couple of D-6 'dozers, each with a pair of grapple jaws, pick up one or more logs at a time then slowly skidded the logs to the central landing where log trucks awaited. The compacted snow and ice on the forest floor made for an easy skid. The cleats on the 'dozers got a good, clean bite in the snow and moved right along the exit path. "Sure, we are working in below freezing weather but unlike summer, there is no dust and no pesky yellowjackets to contend with" announced one logger.

After a few trees are cut and skidded away, the fallers go to work again. This is "selective logging" with only marked trees being cut a few at a time.

Stand of timber in the Hoxie-Griffin Sale, Bureau of Land Management. Sugarpine in center was photographed as it was felled.

One bulldozer, with a long cable and a choker chain, drags the log to within reach of another dozer with a grapple. Instead of dozens of workers cluttering the forest, a crew can be made up of as few as two, maybe three men: A faller, a choker setter and a catskinner ('dozer driver). A choker setter is not always needed.

Protecting the ground and small seedlings is a primary concern on this operation. Often the 'dozer goes out to the end of a skid trail then with blade down, back blades the trail. In this cold weather, the surface freezes then along with more falling snow, the trail becomes ready for use again a few hours later. The bulldozer and the skidded logs never touch the ground.

Daily operations start in the frozen dawn to accomplish the best work before the afternoon thaw sets in. By 1 or perhaps 2 p.m., everything quits until the next dawn. By spring thaw, the logging is planned to be finished and there will be no telltale tracks on the forest floor to show where the bulldozers have been.

When summer comes, it will be doubtful, when the site is viewed from an airplane, that one could tell that over 5,000 trees have been felled and hauled out of this sale in the forest. ◇

About Undercuts

"Undercuts" are the wedge shaped piece cut out of the side of a tree trunk to help control the tree's direction of fall. In the old days of timber falling, the "undercut" was cut out with axes chopping on a "down" angle as shown in sketch at left. The saw cut which felled the tree was then made from the other side of the trunk, cutting toward the apex of the undercut. Once the tree was down, part of the "cleanup" included squaring off the butt end of the log, This cut required an additional full-thickness cut and waste of the end of the log.

These days, the "undercut" is made in an upward direction, wasting only the wedge from the unusable stump and eliminating the second full-thickness cut. This saves the end of the log that was previously wasted. Today the undercut chunks are salvaged by firewood gleaners after the tree falling operation is finished.

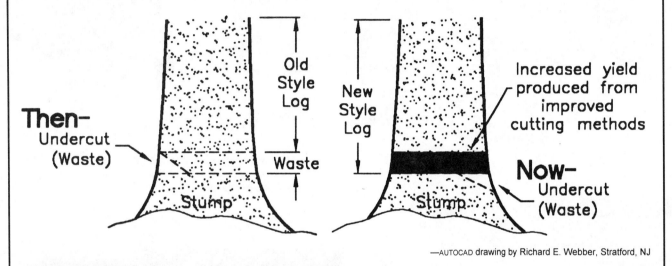

—AUTOCAD drawing by Richard E. Webber, Stratford, NJ

(LEFT) **Historical photograph demonstrates earlier way of chopping out the undercut.** (RIGHT) **Today's method preserves more of the log as well as the undercut.**

27

Snow must be removed from base of tree before felling. Snow here was about three feet deep.

As timber faller starts the undercut, shoveler digs out escape path in the snow through which faller will dash when tree starts to fall.

Faller carefully works on undercut as the direction of fall has been predetermined.

Timber faller lifts undercut away from tree. The undercuts from all the trees are left in the forest to eventually be gleaned as fire wood.

(TOP FROM LEFT) **Working on opposite side of the tree from undercut, faller** (ARROW) **starts final cut.** (RIGHT) **As tree starts to topple, faller** (ARROW) **dashes through escape path.** (LOWER) **Falling tree picks up speed.** (RIGHT) **Tree crashes to soft landing – minimal damage to log – in the snow; no damage to forest floor.**

The sugarpine crashed in the predetermined place in the forest in an effort not to damage younger trees in the vicinity.

The timber faller cut this log into three, 32-foot and one 26-foot lengths before the catskinner arrived to grapple each section and skid each to the landing.

The butt of the log was the first section skidded out of the forest.

Logs slide easily along a skid road of snow and ice. Because of the snow pack, there is no gouging of the forest floor.

Sugarpine log on the way to the landing during logging-in-the-snow project in February 1996.

Dave Northrup, Tree Faller

For the last fifteen years, Dave Northrup has worked in the woods and that fifteen years accounts for nearly all of his adult life. At 34, married and with two children, he owns his home and enjoys a comfortable way of living.

This wiry 185 pounds, six-footer, who lifts weights, is a master of the chain saw.

He works on contract with various timber outfits. The working day runs about 6½ hours. He said that amount of time was for safety and was fairly well agreed throughout the industry.

A faller has to be continuously on the alert – not just for watching the progress with the tree he's cutting, but for what else is going on around you.

The days start early which means being on the site at dawn with the first tree down when it is light enough to be certain where it will fall. Northrup generally stays right with his work taking lunch when the day's job is over.

Pictures on the Hoxie-Griffin Sale were made by Bert Webber, Research Photojournalist, who is shown at the landing.

Dave Northrup, master of the chain saw

Working on the Hoxie Sale [See Chapter "Logging in the Snow"] an early start was necessary in order to get in a day's work because we had to quit when the mid-day weather warmed to where the snow started to get soft.

For energy snack, he might have a granola bar. Snack time generally coincides with a stop to service his gasoline powered saw. In his pocket are a couple of extra chains. A chain can last three or four days with normal wear.

Just like many jobs, spare parts are a must. You can dull or even break a chain by hitting the ground or a rock. If there is just a nick in the saw, it's sometimes better to stop and file it away than to take time to install a new chain.

Dave Northrup said most fallers like to work on the larger trees but these take extra careful work to make certain it falls where the faller wants it. A day's work on big trees, including time to dig snow away from the bases, is about eight trees. On the small trees, a cut of sixty up to 100 trees is considered a good day. Northrup's largest tree was about 7½ feet diameter.

Northrup is a product of the Oregon's Rogue River Valley having graduated from Phoenix High School in 1980. <>

LOG BEING LOADED WEIGHED 24000 POUNDS. SCALED 3900 FEET.

Making a Logging Camp Work

Anton Lausmann incorporated under the name of East Side Logging Company in 1921. His partners were J. P. Miller and his brother Joe Lausmann. Anton considered the most important office in any corporation was that of secretary-treasurer so he took that position. This enabled him to have a continuous up-to-the-minute handle on all the financial arrangements.

The McPhersons were a financial investment company, had organized the First National Bank of Detroit, controlled the Detroit Trust Company and the latter had stands of Oregon timber. One of these was in Columbia County's mountainous region northwest of Portland.

When in Howell, Michigan visiting the Mc Phersons, Anton Lausmann bought nearly all of the McPherson family stumpage around Keasey, just west of Vernonia. The agreement between the McPhersons and Lausmann was that Lausmann would log both the family land and that of the banking syndicate and keep separate books on each. The Michigan management gave Lausmann *carte blanche* on all on-site arrangements.

The land was in Columbia County in the area where Columbia, Clatsop, Washington and Tillamook Counties come together. The business offices would be in Portland's Spaulding Building. East Side Logging Company would cut millions of feet of timber and everyone connected with the operation would make money – lots of money.

Heavy equipment has always been expensive. Anton Lausmann was not a rich man but he had lots of friends who were. He and Miller ventured some money and set up a leasing company. This firm would be named the L. M. Company, Inc. and was independent of the logging operations. The company rented or leased equipment then profited by the interest generated on the notes. One of these deals was an L. M. Company order on Willamette Iron and Steel Works for one 12 x 14 compound-geared two-speed yarding engine. Its weight was 66,000 pounds. This "donkey" engine was ready in early 1927, leased to L. M. Company and shipped to the logging site.

Lausmann was a principal in L. M. Company as well as in East Side Logging Company. He was a young man and he was wheeling and dealing.

For the logging company, he set up his field office at a village, Keasey, about 8 miles west of Vernonia. He maintained his principal office in Portland to have access to dependable communications – telephones, telegraph and the U. S. mail because he needed to be in firm contact with his mid-west and eastern buyers.

There had been a rural post office in a farm house at Keasey but this had been closed. An early concern of Lausmann's was the lack of postal service to his field operation at Keasey. It didn't make business sense to run back and forth the eight miles to Vernonia for mail as there was no road, just a rough trail and a railroad track.

Tony Lausmann hit upon an idea that if he built a suitable post office structure at Keasey then pressured, in a friendly, business-like way the postal authorities, he didn't see any reason he couldn't have his postoffice at the Keasey switch yard. He assertively pursued the idea. When the

One of a number of donkey engines at East Side Logging Company.

Railroad interchange at Keasey, Oregon in early 1920's.
All buildings and rails were later removed.

U. S. Post Office Department argued that they didn't have a place to establish a post office, Lausmann declared, "I just constructed a new building for this purpose and all you have to do is move in to it." The Oregon-American and Inman-Polson timber operations also switched out of Keasey so there seemed enough business to justify the post office. The Post Office Department accepted the offer.

At this time telephones at Keasey were unknown so as he set up his Keasey operations, one of the first considerations was to install his own private telephone system. Lausmann had done this elsewhere, earlier, and with great success. Once his system was in, he petitioned the local telephone company to provide a trunk line to his central exchange. These special lines were costly to get and sometimes this meant the telephone company had to run miles of wire for the hookup. But Tony Lausmann was not a piker when it came to paying for services wanted. With the line up, his Portland and Keasey offices could talk.

Tony Lausmann often played a concertina at union meetings to break up the tension. Shown at age 83, he was playing for his neighbors.

Further, with Tony's home-made switchboard, he, in Portland, could talk to any of his sub-offices in the woods back of Keasey.

Lausmann bought an old house that was near the railroad switchyard. It had served as a community store, post office, and on the upper floor there were rooms to rent with breakfast thrown in – a Bed & Breakfast in the 1920's. He occupied the building as a store and kept the upstairs set up for overnight guests there being no other facility at Keasey.

Because the population of Keasey was growing, the trail into Vernonia was graveled which now meant that automobiles, mostly Model-T Ford's, could get to Keasey. Men who worked for Inman-Polson as well as Oregon-American were parking their cars in the woods around the switch yard which caused congestion. Lausmann observed this and decided there was money to be made. He bought corrugated iron and built about thirty garages. He reserved several for East Side Logging Company, then watched as all the others were promptly rented. This was another of his

Anton A. "Tony" Lausmann was nicknamed the "Swivel-Chair Logger" because he spent his career in the office making deals. He did not work   in the woods – never cut a tree.

private money-making ventures.

He was committed to the idea of never keeping his money tied up in one enterprise. Each operated independently from the others but were inter-related for policy and sources of supply. All the bookkeeping was done in a single office – his. He was constantly on the lookout for new ventures or ways to improve his established businesses which were all corporations. One of his techniques was to engage in ventures that would serve his other interests. This way, there was pass-through money and each business benefited. Of these enterprises, East Side Logging Company was the largest and took most of his time.

Tony Lausmann had been brought up in Chicago. He had gone through the 8th grade, attended a business school where he studied bookkeeping, and he learned to be a manager. Managers, he had been taught and later observed, were men of authority and they wore shirts with neckties. He was a gentleman – not a rough-and-tumble logger. But when he was in the woods, he wore hob-nailed boots. Lausmann was not tall, but he walked with unquestioned authority.

Punctuality was his middle name. Visitors to his offices, usually salesmen, had to have advance appointments. He was of the firm belief that time was money and he did not want people to have to wait for him and he did not want to be kept waiting when he went to see someone at their appointed time.

When there was a chore to be done, he did it himself if it was accounting. If it concerned operations, he delegated it to a field superintendent.

When it came time to hire men to work in the woods, he left this to his camp manager. He had done hiring earlier, when he was the bookkeeper at Stanley-Smith Lumber Company at Hood River, so he knew what to look for.

Re recalled this activity and related them to the author:

Very seldom did a man wander into the General Offices in Portland and say he needed a job. I very seldom came in contact with anyone that wanted a job for the men knew they should go to an employment office or to the camp. But if we needed office personnel, I handled that.

Because the camp was in the mountains out of Keasey and hard to get to, most of our hires were at the agencies.

For years we used the Red Cross Employment Agency. It was on Front Street in the flop house and skid road area. When one of my superintendents needed a man for a special job, a blacksmith or such, the manager would specify this to the agencies. When a suitable applicant presented himself, the superintendent in the woods would get a telephone call. If the fellow seemed satisfactory, he would be given an employment card and sent to Keasey on the train as most men did not have their own transportation. The possibility of a man being hired by merely wandering into the camp was not very good. For one thing, I was firm with my superintendents that there be no loitering in the camp. We hired a lot of men. Sometimes if there was a big need and a contingent was to go to camp, someone from the woods went into Portland to meet them and escort the group to the camp.

In Columbia County, Tony had two separate timber operations running at the same time. In addition to East Side Logging Company out of Keasey, Lausmann had Rock Creek Logging Company that was about in the middle between Vernonia and Keasey. The United Railways served this area. With Oregon-American, Inman Polson and East Side as well as Rock Creek all employing men, it had to be very clear where the workers would get off the trains. To leave no doubt where this was for the Rock Creek works, Tony arranged that the platform (station) for Rock Creek Logging Company be named "Lausmann Station."

The name "Rock Creek" could not be used as that name was too similar to stations elsewhere. Knowing when to get off the train for the East Side camp was easy as that was at the end of the line. It was important to be exact in the directions for many of the men did not read and often they were "under the weather due to 'bottle' problems," Tony pointed out.

At times there was no choice in the quality of workers available for general labor. Lausmann recalled:

We had to take what we could get. Sometimes these men were dissipated and could not stand up physically to the demands of the work. When this happened, all we could do was to pay them off and send them back to town.

By agreement with the railroad, East Side Logging Company could issue travel vouchers good for a one-way ticket between Portland and Keasey or in the opposite direction. This took care of a few men who needed to be moved to the camp for work and were penniless or, some who had been fired and were "shipped out on the first

Bunkhouses at Rock Creek Logging Company. The raised plank sidewalks were needed to help preserve the bunkhouse floors from the mud that in spring was a nearly-daily hazard.

train" to get rid of them.

When it came to the more skilled job specifications, the employment offices screened the applicants closely to be certain the men could do the work. This was because the reputation of the agency was at stake. The agencies knew the applicants' previous record of work and the extent of capability each had established in the work place. It was to the advantage of all the timber companies to help the agencies by answering inquiries of the agencies as to how good a man really was with his work, his reputation, and if he quit or was fired, exactly why.

Some agencies were pretty loose and got jobs for men without looking for references.

Lausmann admonished:

The hiring halls might get away with sloppy placement with unskilled workers, but seldom when a specialized job was at stake. When work was scarce, an agency had to be pretty careful whom they sent. When an agency became careless, the word got around to the various operators and that agency ran the risk of not being called.

It cost a lot of money to put a worker on the payroll. We had to issue a lot of stuff the men would have to work with, feed him, provide a bunk then discover he couldn't do the job. We had men coming to us from all over the nation wanting to work in the northwest woods.

In the summer, men with excellent educational backgrounds, who had winter jobs, came to the woods to work. Many were school teachers who had been cooped up in classrooms. For them, a job in a logging camp was like a paid vacation.

The work refreshed them, brightened their outlook on life and readied them for the classroom in the fall. There was seldom any trouble with these seasonals. Many came back year after year and their academic experience often rubbed off on some of the "under-educated" men.

Everyone who passed the basic employment agency questionnaire was given a chance as a general laborer. There were a lot of college boys in the summer but too often, these fellows truly did not want to work and it did not take the foremen long to find it out. If a fellow was a "goldbrick" (goof-off), then the quicker rid of him the better if from no other standpoint than safety.

Lausmann:

If a fellow was sloppy in his work, he could be a safety hazard. We just did not have room for careless workers. These part-timers were mostly in the summer and this was fire season in the forests. We could not take chances.

The superintendents, as Lausmann called his field managers, had great responsibility. In the early days, the managers and the foremen appointed under them, had to be men who know their management responsibilities as well as knew the details of all the other jobs. They also had to be bigger, stronger and able to throw their weight around when necessary.

Wood's crew, just in from a day's work, wait to be photographed before eating dinner. Dinner, the major meal of the day, usually taken at noon, was served in the evening at the logging camps.

Lausmann:

Years later, when the industry grew more refined and machinery became sophisticated, a superintendent with working brains was more important than one with mere brawn.

Tony Lausmann's school training had been in accounting and management but he accumulated much knowledge of camp operations in previous employment at Hood River. Since the purchasing of equipment for the mills was his responsibility, he read a lot, listened carefully to salesmen and visited other mills. There was a sort of fraternity among the management of most timber outfits and when a question came up from among the "fraternity," it was discussed. After all, there was plenty of stumpage available to cut so why not be friendly? He talked with many men who had been in business longer than he.

In the camps there were time keepers, bookkeepers and commissary men who would be interviewed for their jobs by Lausmann himself. In addition:

I audited the work of everyone of these paperwork men regularly since poor accounting could cost me my shirt.

Tony talked with the camp cooks about prospective cooks who had applied for work by telephone. He discussed kitchen equipment as well as recipes as it was the Chief Cook's responsibility to buy the food. Before hiring a cook, he was certain to learn if the man was a general cook or a baker only. He had learned about this the hard way for at an earlier camp he hired a "cook" who turned out to be only a dessert baker. There was room for bakers in his camps because of two factors:

1.) It was less costly to bake bread in his field kitchens than to have commercial bakeries in Portland try to ship bread to the camps in a timely manner.

2.) The loggers liked what they called "hardy" bread, not the soft, no-flavor, pure white factory-made loaves.

Lausmann said he wanted to get the right man in the right spot.

We didn't want a baker when we needed a beef stew man. The bake shops were usually separate from the main kitchens as some bakers were cranky about their craft and some would not talk with the meat-and-potato cooks.

The bakery was busy because bustling bakers baked beans, beautiful buttery banana buns, blueberry biscuits as well as cakes and pies.

As a rule, the Chief Cook was the boss of the cook house and the baker was accountable to him. It was sometimes a standoff however as the bakers wouldn't talk with the cooks therefore the cooks would not talk with the bakers.

It was an urgent matter that we hire good cooks because if the food didn't suit the loggers, they would quit and move to a camp with better chow. It didn't pay to buy low quality food for the same reason so we bought the best available.

Accordingly, with good food products to work with, the cooks and bakers had to turn out top stuff or we'd go look for a new kitchen crew. These workers all knew this.

The largest number of men served in a Lausmann dining room was 130, and they had to be fed sit down meals twice a day. The meals were very hearty. Our loggers got all they wanted to eat and it certainly was not corn flakes and skin milk for breakfast. We had cereals around, but the favorite breakfast would be lots of eggs and bacon, hot cakes and syrup and of course coffee by the gallon. Fresh milk was hard to come by because we were so far from a source and this was before there was refrigeration. So we used canned milk. You had to add water to it before most men could drink it, but a few took it straight out of the can.

Punching holes in the tops of the canned milk was never any trouble as long as there was a nail and a hammer around. In those days the design of the milk can would not fit in a can opener and this was long before the invention of the most famous tool of all time, the "church key" (beer can opener).

We had one cook who considered himself quite an ax-man. He prided himself in getting a hot fire going very fast in the cook-stove because of his particular manner of thin-splitting wood with a hatchet. Once, when he couldn't find a nail to open a milk can, he tapped the can with the corner of his hatchet and thus we had canned milk splattered all over the ceiling.

The eight-hour day didn't come along until after the "wobblies" (Industrial Workers of the World – I.W.W.) came into the picture. Before then, the ten-hour day was the practice. In the logging camps, breakfast was between 5 and 6

There were two old box cars that had been fitted up as crew cars, complete with a toilet booth in each car. These old "closets" came from wornout passenger cars. On each was a sign:

DO NOT FLUSH THE CLOSET
WHEN THE TRAIN IS STANDING
IN THE STATION

(This was long before the invention of self-contained "Porta-Potties.")

a.m. The men worked quite a distance from the camp and each was issued a "dinner bucket." On the way out of the dining room after breakfast, they would stop by a table that was setup with cold cuts and other lunch type foods. Each man would fill his bucket then took it into the woods with him. This was a lot of work for the cooks because they had to work harder in the early morning hours than for the later meal because they had to set out breakfast as well as the lunch table. Breakfasts were always hot and the lunch stuffs were always cold, even in cold and wet weather. It often rained in the northwest woods.

To bring the men in for lunch would have been too costly in down time as the trains only moved between six and twelve miles an hour.

Many times efforts were made to get hot soup into the forest on very nasty days but this was never successful.

The "dinner bucket" was indeed a bucket with a bail. At first these were made of sheet metal about the thickness of a tin can. They were hard to keep clean, dented easily and would rust. Later, more durable models were aluminum. Each had a flat cover with a small handle in the center for lifting the lid from the top of the bucket. The handle stuck above the surface of the lid making the lid useless as a flat plate for when putting the lid with its top side down on a stump – the usual dinner table in the woods – it would tilt. Some fellow got the idea of placing a couple of handfuls of dirt on the stump then burying the lid's handle

Taking a break during the day's hard, rough work was a natural thing to do even if the coffee was cold.

in the dirt thus steadying the lid so it would be usable as a plate.

There were a lot of complaints about those lids to a point where Lausmann contracted with a Portland metal shop to custom make lids with countersunk handles. (It would not be until the start of World War-II that lids for mess kits, "dinner buckets," had a groove down the middle for the handle and deep dishes on each side of the handle thus allowing for liquid foods to be served in the tops of the mess kits.)

The men took coffee into the woods if they furnished their own containers. Glass-lined thermos bottles were a lost cause as they broke the first time a bump came along or if a thermos was dropped. Army canteens were very good and a lot of them men had them. Once metal thermos bottles came on the market, Tony Lausmann bought dozens of them and issued one to each man. But hot coffee put into the carriers in the morning was always cold by mid-morning. A firm rule strictly enforced was that no man could build a fire in the woods, not even a fire small enough to heat his canteen of coffee. When miniature Sterno* stoves came along, these were also banned because the canned fuel emitted an open flame.

Come noon, there were always a dozen or so, like the men from the office, workers in the repair shops, saw filers and maybe a train crew, who took lunch in the dining room. This meal, although served at the tables, was the same fare the men had taken in their buckets into the forest – cold cuts. But the coffee was hot.

On weekends, many of the men went to town and stayed overnight. They had to be back on the Sunday evening train. There never was a grand exodus for town for many of the men didn't have the money to carouse so they stayed in camp. In the cookhouse, this meant three sit-down meals had to be served on Saturdays and Sundays. Because there was not generally any work in the woods, the cooks did not put out the heavy meals served during the week. On Sundays, breakfast was not until 8 in the morning. This meal might be fried thickly-sliced bacon, fried sliced potatoes with bacon gravy and the usual black coffee. The bakers always shined brightly on Sunday morn-

* Sterno was solidified alcohol in little cans about 2-inches in diameter and 2-inches high. Sterno was also banned from the camp because skid road types were known to squeeze-strain it through a sock and drink it.

The Chicken Dinner

The 4th-of-July was not a normal holiday in the woods and of course fire works were prohibited. But for the evening meal, by planning months in advance, a deal was struck with a farmer near Vernonia to raise a large flock of chickens for a fried chicken feast. The going rate for a frying chicken was 20¢ each "on the hoof." But the cooks had no way of slaughtering 80 chickens so the superintendent allowed an extra 2¢ apiece if the farmer would ring all the necks and pluck the birds then put the chickens in gunny sacks for delivery to the cookhouse. After dinner, and at dusk, in a clearing, where a lot of the nearby slash had been stacked, there was a huge bon fire.

Lausmann said, "Come to think of it, we didn't have a single American flag in the camp."

ings for this was the one time of the week when they had time to prepare and serve savory special sugared coffee-cakes. The head-count for Sunday meals was about half the usual crowd, the number to be fed determined late on Saturday night.

Every camp had its challenges concerning drinking water. This was especially true in hot weather. For some reason known only to the U. S. Army, a soldier was supposed to be amply supplied with his single canteen of water for a day in the field. But the Army's reasoning didn't work in the woods. Although the work train carried drinking water and there was always a ten gallon milk can of water near every donkey engine, this did not help the men who worked far into the forest. In hot weather, it was not uncommon to see nearly every man leave camp for a day's work carrying a gallon-size jug full of drinking water.

Earthenware "moon-shine" jugs were prevalent and would withstand more knocking around than glass. Plastic jugs had not yet been invented.

Lausmann added:

On farms, ranches and in town, the main meal of the day was usually at noon. Of course this was not practical in the camps where the majority of the men had their mid-day meal out of their buckets. Our big meal of the day was served at night. During the 10-hour day-era, this could be as late as 7 p.m. For evening dinner we had most everything that was available.

We bought most of the fruit and vegetables from Pacific Fruit and Produce Company. Mason and Erman along with Wadhams & Co. were our general grocers.

Our meats came from Armour or Swift packing houses. I bought beef by the whole and had it delivered in quarters. The cooks would hang the beef in the meat house where we had the chopping block. We did not hire butchers. The cooks prepared the meat into cuts that suited the menu for the day. This was often steaks and of course we had a lot of stews and boiled beef roasts.

The camp cooks used potatoes by the bushel and on delivery stored them under the cook house floor where it was cool and dark. Boiled potatoes were a regular item and quite often there were baked potatoes, especially on Sundays.

Loggers seemed to have a "thing" about vegetables. There were few they would eat. Boiled carrots yes, but never raw carrots. Boiled cabbage, yes, and cole-slaw just a little. They ate canned peas if mashed potatoes were on the same menu so the men could mix the peas with the spuds. But never peas alone. The men shied from ordinary string beans but of the coarse and tough "leather britches" variety, they ate all that were served.

The men loved dumplings and it takes a good cook to make a good dumpling. But this didn't happen too often because of the extra work. From a management standpoint, it was a tossup on costs between the price of spuds boiled in a big pot with almost no labor, and the labor-intensive hand making the dumplings which was little more than flour, soda and water. Home-style dumplings might be on the order of a golf ball. In the camps, the minimum size was tennis ball size but larger was often the outcome.

The dining rooms were usually in the same building as the kitchens. The combination building was just called the "cook house." Tables, of sturdy construction, were set with hotel supply house heavy-duty china. In some of the camps that Lausmann bought out, the cook houses were equipped with enamalized iron dishes. If this enamelware was dropped, it would chip every time. Then a few days later rust appeared. As fast as these plates and cups were damaged, they were thrown out and replaced with the heavy hotel china.

In the big camp like East Side Logging Company, there was a Chief Cook, two and sometimes three second cooks plus several bakers. In addition, there was a hired staff of "flunkies." These were the waiters and the people who cleaned up after the meals and washed the dishes and scrubbed the floors.

No Lausmann outfit ever adopted the army stunt of pulling a high-priced man off his special duty to take a turn in the kitchen washing dishes

Quality and Quantity

Logging camp food was graded slightly better than the food in the average homes. If the men were to be kept in good spirits, the best way to do this was through their stomachs with good, substantial, tasty meals. The cooks were, by far, the most important employees.

One morning I did a surprise inspection by arriving at breakfast time. I remember once when I went into a cookhouse and the cook complained to me that a logger took an entire platter of eight fried eggs and dumped them all onto his plate. I smiled, then directed the cook to fry another batch. It was true, the camp that served the best chow got the best men. It was a simple as that.

Work stopped almost like magic when a traveling photographer appeared in the woods or in camp.

for a day. In addition, the rights of every man were respected and no penalties were assigned to men as was done in the army by top sergeants. Lausmann never forced a man from the woods to scrub pots for some infraction. If a man was causing trouble in the woods or in the bunkhouse, he was talked to by the foreman or sometimes by the superintendent. The man had a choice. Either straighten up and follow the camp and work rules, or he was terminated and shipped back to town. If he got canned, the hiring halls in Portland were quickly told.

In the early days, Lausmann hired women cooks and waitresses but he found this was detrimental to everyone's best interests so he switched to all male kitchen workers. Unless the cookhouse was operated by a contractor who had his own rules, the Lausmann rule was no talking at the dinner table except to ask for something to be passed.

With many dozens of men chattering away along with the clatter from the dishes and utensils, the racket could become un-bearable. In addition, if there was no talking, there would be no arguments. Of course anyone could speak to a flunky about the food – we were not that strict. But we once had a fellow who insisted on giving back talk to a red-headed waitress (flunky). She took a platter of food, hit the guy over the head with it and nearly put him out for good.

We got the opinion from the other men who saw the incident that the men figured that his girl had been provoked so we canned the man. As the girl was a good waitress, we kept her.

During the days when talking was permitted, arguments often led to fights. Even after the ban was imposed, fights continued but to a lesser degree. The men knew the 'why' of the no talking rule and most of them went along with it.

The loggers were together for weeks at a time and the work was strenuous. They needed some entertainment but there was not much to do in the camps. There were cards and dominoes and in one camp, eventually slot machines. Some of the men lost their tempers over card games and once there was a knock-down, drag-out fight. The superintendents did not consider a fight to be serious and demanding of management's attention unless the men were so beat up they could not do a full day's work on the morrow.

Some camps canned men promptly for fighting but those camps soon had trouble getting anybody because it seemed that most loggers had to

THE RULES
Of
CRIBBAGE
The Ever Popular
Card Game

Established HORN 1846

W. C. HORN, BRO. & CO.
Manufacturers Since 1846
NATIONAL HEADQUARTERS FOR ADULT GAMES
Salesrooms   Factory
Fifth Ave. Bldg.,   571 North 3rd St.
New York   Newark, N. J.

get into a scrap once in a while just to let off steam.

One of the popular entertainments, other than cards with its attached gambling, was arm wrestling. In some camps, the long summer evenings were spent in arm wrestling tournaments. Often there was money changing hands after a match.

Card playing was usually poker, black jack or hearts – always for money. Typically, games started as penny-ante stakes but as hours passed the piles got bigger. The only thing that saved some paychecks was the mandatory lights out hour.

There were a few chess players but these, and some cribbage players, were in the minority except in the summer months when the school teacher-loggers were in camp. Most of the teachers played for fun and often had tournaments. Also, many of the school-types needed the money back in town so did not gamble at all.

We always kept the company store open in the evening hours and on weekends. This was handy for the card player-gamblers for many men established rules that a new deck of cards was required once an hour. The gamblers insisted that the new decks had to have a manufacturer's seal on the packs when purchased. They usually burned the used decks to 'get rid of bad luck' after a particularly strenuous game.

An excess amount of bookkeeping was being created at the company store due to the huge numbers of "chits" the men signed for purchases. We did not allow much money to be kept by the men due to the risk of theft. If a fellow wanted a 5¢ pack of cigarettes or a bottle of soda pop, he would sign a chit. A bookkeeper at the camp suggested that East Side Logging Company coin "company money." This way a man could sign for so much in dollars of the "funny money' against his pay check. The special coins were good at the company store, the gamblers accepted it, and it eliminated the paperwork with all those chits.

A man could always turn in his chits for hard coin-of-the-realm at the company office in Portland, and if a man was terminated for some cause, his coins were redeemed by the camp superintendent before he was handed his walking papers. The private coinage idea was new at the East Side Logging Company but many timber operators used it.

Lausmann:

We decided to try it. A metal stamping company created the design and provided denominations of nickels, dimes, quarters half-dollars and dollars in aluminum slugs. These were actual money size but much lighter. One side had the name of the company EAST SIDE LOGGING COMPANY set in a circle with my initials AAL in the center. On the reverse in a circle was the phrase, GOOD FOR at the top and IN TRADE at the bottom. The denomination was a large number in the center.

It was never intended that this "funny money" as the men called it, be good with merchants in town. Nevertheless, one day an outfitter in Portland telephoned the office saying that a clerk had accepted several dollars of these company slugs because he recognized the company name. Would we redeem it? I told him to bring it over and we would give him a check for it.

The camp bookkeeper made another suggestion and Tony Lausmann quietly jumped at it.

Put in some one-arm bandit slot machines for entertainment of the men and the profits of the machines should pay all of the costs of the company's funny-money project. The machines would be tuned to accept the company slugs as well as regular United States coins.

Tony Lausmann bought slot machines. While he was at it, he put in a couple of Nevada gambling tables. The arguments over table gambling brought more fist fights than ever before so Tony dumped the tables.

It was surely true that whenever a bunch of single men are sitting around talking – it does not matter what the subject was to start with – more than usually the subject changes to women. In a hardy bunch of loggers, the chat often drifted to the number of "conquests" a given fellow had made on his last weekend in Portland. The language was pretty crude and sometimes just plain vulgar. When an expletive was emitted within the hearing of a waitress, some other fellow might demand that the first man take it all back and apologize to the lady. As apologies were seldom forthcoming, these encounters usually ended in another fight. Occasionally these fights became a brawl of several men in a dog-pile. It was not often that a free-swinging free-for-all developed, but when this happened, about the only way to bring order was for a superintendent to douse the bunch with a fire hose.

Although hard liquor was forbidden in the logging camps, some got in with every bunch from a weekend in town. The more serious fights

To ease bookkeeping due to hundreds of "chits" loggers signed at the company store, company money was issued. Loggers called it "funny money."

usually came within a couple of days after the liquor ran out and the chronic drinkers got bored.

The policy about drinkers varied. If the fellow was a good worker, he was called in for a "chat" (counseling) and was told if it happened again he would probably be terminated.

There was no personal "inspection" of the men when they came back from town. Some fellows, who came back to camp roaring drunk and were absolutely obnoxious, were just canned on the spot, not admitted to the bunkhouse and were sent back to town on the train's return trip. On the other hand, some returned to camp in fairly good shape – maybe a little rickety – but not of the brawling type. These slept it off, had a little hangover the next morning that a good breakfast and lots of coffee seemed to straighten out. To these men, a degree of tolerance was permitted and in many instances the camp superintendent never learned about it other than by hearsay.

Although logging trains were far more common in appearance at the Keasey switch yard, there was regular passenger service to Portland by way of Vernonia. When the Sunday evening train arrived, speeders from the three logging outfits were at the interchange to pick up the men. The crew cars for East Side were also used, each car taking about twenty men and their luggage. The luggage was usually pretty light. The distance to the camp, which was at the top of the divide, was about three miles. A ride on the speeder or in a crew car behind one of the locomotives was a lot better than having to hoof it with a suitcase, especially on a chilly night, if one had his snoot partially full of booze.

Lausmann:

We tried to keep the number of rules in camp to a minimum because the more rules you post, the more policing the foreman would have to do. We were a profit making venture and men were hired for just one reason: work. If there were constant troublemakers and/or men who would not do a full day's work, we canned them.

In some camps, kerosene lamps were all the light available as there was no electricity. In others, we had power but we put the lights out at nine o'clock. We did not have to enforce the lights out. It was just accepted. The majority of the men realized if they did not get all the sleep they could, they would not be worth anything the next day in the woods.

Lumbering has always been a strenuous job and it remains so today even with chain saws instead of a long, hard-to-handle hand buck saw.

Probably the most precision job in camp was that of the saw filer. In the early days, this work was done with a file by hand, carefully but firmly applied to the teeth of the saw to file the edges of the teeth in as sharp a profile as possible. The hand-filers took great pride in their work and while they were at it, they inspected the saw carefully looking for any cracks that could cause the saw to break when in use. Great safety measures were required in the "saw shack" and loitering in the shack was prohibited.

It was traditional that each bucker and faller provided his own saw. These were the days of elbow grease and sweat, long before gasoline-driven chain saws. There was no schedule for sharpening saws, it was just done when it was needed. When the train brought the loggers out of the woods at the end of a day, it was not uncommon for the filers to have to stay up half the night to have the saws ready for the nest day's work. The men were careful with their expensive Distons and to use another man's saw without permission could spark a fight.

The one thing that would cause a sawyer to cuss louder than the snorting of the donkey engine, would be when he hit a piece of metal in a log. Many times in raising cables, spikes would be driven into a tree. When the job was finished, the cables would be taken down but the spikes remained to be grown over in the following years. Eventually, when the spiked trees were cut, the unknown hazard – the buried spikes – could be hit by the saw. This would instantly dull the teeth and sometimes ruin a saw. After the advent of the preservationist and environmentalist era, spikes would sometimes be purposely driven into trees as a protest against cutting timber. When a high-speed gasoline-powered chain saw hit a spike, the chain could easily break, fly through the air and risk hitting and injuring nearby loggers which seemed to be the idea as such accidents might shut down the operation.

There was one man at East Side Logging Company who was called a "bull cook." But he

The railroad, cookhouse, bunkhouses, had been built at the Divide Camp at East Side Logging Company and the business of logging was about to begin.

didn't cook anything. His job was to cut the fire wood, buck it into stove lengths, clean the bunk houses and when bedding was furnished, be made the beds. He also kept the four-hole privies clean.

The "flunky" worked the dining room and the kitchen and did not work in the bunk house. This latter was a hangover from the days when women were the flunkies and women were taboo in the bunk houses.

Lausmann:

The first time I was in a bunk house the beds were wooden platforms – three deckers – and each deck was large enough for three men across. At that time, all that was furnished by the operator was a space for the man on the plank, and some straw. We had no way to keep the straw from drying out so the stuff kept falling down into the eyes of the men on the lower decks. Each man had his own blankets. He would curl up in his space on the tier. There would be a maximum of nine men to a unit. It was very difficult for the fellow who had the top level, inside space to get in and out of the rack.

The cooks did not serve boiled or baked beans very often for when they did, the air became so "be-fouled" during the night one could hardly breath in the stench-filled building. Men in the bunk houses did not often open the few windows for ventilation, if the windows opened at all.

When the unions came in, one of their first moves was to get rid of the platform bunks. Steel cots were bought but the straw was continued which was put in bags similar to mattress-covers. Eventually, under the cleanliness rules of the unions, the straw gave way to mattresses, sheets, pillows and pillow cases. Because all these beds had to be made daily, the operators were forced to hire more bull cooks.

At times, bull cooks sometimes presented trouble for themselves as, inevitably, some turned out to be homosexuals. Most loggers were rough and tumble men who were proud of their maleness. These men would not tolerate a "queer" in their midst. When a homo was discovered in a bunk house, he was gotten rid of very quickly, usually, unfortunately, after the men had beaten him up.

Another troublesome matter concerned women in the camp. When women were working in the cook house, they were supposed to be restricted to that building. They had living quarters either in a room within the building, or in an adjoining shack. But at night, those interested in extra "work," didn't always stay in their own beds. A real measurement for discontinuing the women flunkies had to do with their night prowling.

49

Lausmann:

Some one or more of the waitresses at East Side Logging Company picked up a case of the clap some place and spread it all through the bunk house. In no time at all, we almost had a complete shut down.

You often hear of these things but it's something else when it happens in your own camp. Years earlier, at Green Point, I observed a gal come in to get her pay check. She pulled down her sox and made a point to show me how many bills she had, how much money she had collected. She made more on the side that she had coming on the payroll.

From an management standpoint, Lausmann left the policing and rules of the camp up to his superintendents. Some argued that to have a couple of women around would take the strain off the boys so the boys would work better.

Even after we officially discontinued hiring women, I don't think it ever got to the point where we were able to keep women out of the camp entirely. I insisted on high moral standards in our camps but problems did crop up now and then. But the women-caused troubles in our camps were very few after we got rid of women in the cook house.

It took all kinds of men to run a logging camp. To realize that the hiring and firing was done without master-degree holding "personnel directors," or "Human Resource Management Specialists" is, by today's standards, no less than amazing. But the old system worked admirably. Between Anton's interviews for the superintendent and foreman jobs, and for his field accountants, the collective heads of the superintendents and foremen for hiring work crews, all the positions were filled and the work usually ran smoothly.

A logging comp, often so many miles distant from any city, let alone even a village, was a community in itself. Every requirement for life had to be available. All this was costly to the management who had to provide the necessities of life but did not have to give frills. Prior to 1909, when corporate income tax was introduced, there was no "write-off" for the costs of doing business.

Bath tubs and showers were nonexistent but there were wash tubs, lakes and creeks. On a hot summer day, it was common for loggers to strip down and flop in a creek to cool off during the lunch break. As Tony quipped, "Hell – the fish didn't mind and there were no environmentalists around."

In those early days, the camps provided the wash tubs and bars of Fels-Naptha soap so the men could do their laundry. The camp also provided wash boards. The men furnished their own rope on which to hang their clothing to dry. If there were cloths pins, the men brought these. Only a few men took regular baths and when they did, the wash tubs, and the Fels-Naptha soup worked very well.

With the advent of sheets and pillow cases in camp, there was a weekly bed-stripping schedule. On that given day, the men would put their dirty sheets and pillow cases in boxes on their way to breakfast. The dirty sheets were sent by train to Portland to a commercial laundry. Personal laundry was not allowed.

We had to have twice the number of sheets as we had beds as one set was on the beds while the other was in the wash. At first we tried to do as housewives do, rotate the top sheet to the bottom then only having one sheet per bed per week to launder. But in the bunk house that didn't work as we could not dictate how often the men took a bath. At times, many of the men smelled pretty rank so there wasn't any use in just changing one sheet at a time for each bed.

One spring, when the bed linen program was starting, we tried getting the men to take baths on the days when their beds were changed. The various bunk houses had different bed-changing days to spread the laundry load over the week and besides, we didn't have that many wash tubs for all the crew to bath on the same night. But we never could convince most of the men to take baths.

We had to keep a pretty good stock of spare sheets for as the program moved along with time, the older sheets wore out. Since all of the sheets had been new at the same time, most of them wore out at the same time. But we didn't waste the tired sheets – tore them into rags for use in the shops.

In the early days, the men carried their own bed rolls and had sheets only if they provided them. Most didn't. In a bunk house in the early spring or late fall, it was cold and one could count on the wood stoves going out during the night. Even so, anybody that wore a nightgown was a sissy but knit sleeping caps were common.

Eventually, the wood-burning bunk house stoves were replaced with drip oil stoves. Lausmann determined he could get the men off to a better morning's start if the quarters were warm.

All of the Lausmann operations were established for only one purpose – to make money. He believed in spending a buck to make a buck but he was careful how the money was spent. He sought markets for his graded logs throughout the nation and until the bottom fell out of the market during the Great Depression, he was eminently successful. *◇

When fire got a start either in or near a logging operation, it was part of every man's job to turn out to help fight it. These were days long before air-tankers with fire retardant, even bull dozers, therefore fighting was hard and often without results.

---

*Anton Lausmann lived to become a retired timber baron octogenarian. During his lifetime, he created and worked many timber and lumber manufacturing operations including Kogap Manufacturing Company in Medford, Oregon. The acronym KOGAP stands for "Keep Oregon Green and Productive." Lausmann started the first privately financed tree farm in Oregon to make certain that Oregon would never grow out of trees.

Railroading in the Woods

One of the great hurdles in setting up a railroad logging outfit was in gaining right-of-way for track that had to cross adjoining private property. For East Side Logging Company, there would be no delays since the track would run over McPherson Syndicate or McPherson family-owned lands.

When in Howell, Michigan visiting the Mc Phersons, Anton Lausmann bought nearly all of the McPherson family stumpage around Keasey, just west of Vernonia. This private land abutted the syndicate land therefore there was no opposition to laying rails across the properties.

With the timber cutting contracts in hand, Lausmann had no difficulty getting financing. In addition to needing a complete forest camp and all its buildings and equipment, he would also have to build a railroad from scratch and arrange for all the rolling stock. He turned to the traditionally friendly junk dealers and set them scrounging for used equipment. What he couldn't get used, he would have to buy new.

Lausmann:

My brother Bob was instrumental in designing new equipment and in redesigning standard catalog equipment. We'd submit our ideas to a manufacturer and in due course, received a shipment of equipment just the way we wanted it. We changed specifications. We had the steam pressure raised on donkey engines, some of which were built at the Washington Iron Works in Renton, Washington. We talked with Willamette Iron and Steel Works in Portland and had them build machinery for us. When it came to wire rope (cable), we were after used rope – miles of it.

As we have seen, he hand-picked his management team – superintendents. They had to be first of all, honest and next, knowledgeable.

We could not risk foolish mistakes or damage to equipment because of carelessness, waste, and lack of knowledge. If we had losses in the operation it could cost us down-time and when we're not working, we are losing money.

Willamette Iron and Steel Works locomotive factory No 25. Only two were built. This is East Side Logging Company Road No. 107. The logging company took delivery on September 12, 1926. It was sold in 1933 when the firm went out of business during the Great Depression.

There were several major railroads in Portland serving Oregon. Of these, the Spokane, Portland & Seattle (S. P. & S.) Railway had track west from Portland along the Columbia River to the coast. A short line, the United Railways, connected with S. P. & S. at Linnton and wandered northerly through the valleys to Vernonia then to Keasey, 9 miles further to the west. In all, the United Railways had 52.231 miles of 70 and 90 pound main line rail. The line had been completed at the end of March in 1923, just in time for Anton Lausmann to decide to connect with it.

East Side Logging Company was incorporated on January 20, 1921 but did not operate until 1923. It spiked down its rails to the end of the United Railway trackage extending railroading for its logging trains 22 miles into the mountains to its camps and loading sites. It used 60 pound rail. The "interchange" as it was called, at the end of the main line, which was about one mile from the Clatsop county line, was a switch yard in which several logging outfits had pooled their tracks. Inman-Poulsen, Oregon-American and East Side each had a trackage there. *

(In 1976, when the author visited the Keasey, one would never know that a major switch yard had once occupied the place. Where all the tracks had been was only a big vacant lot with weeds and scrub brush.)

Tony Lausmann recalled that East Side and Oregon-American used disconnected trucks. Inman-Poulsen Lumber Company shipped all of its logs on either regular 40-foot long logging cars or flatcars.

Although there were many brands of gear-driven engines especially built for the logging industry, Anton Lausmann preferred the Shay. This was a product of Lima Locomotive Works in Ohio. Because the patents on the Shay went into the public domain, a geared locomotive built by Willamette Iron and Steel Works in Portland was a near-copy. He owned four Lima Shays over the

* Inman-Poulsen Logging Company is listed in the *Encyclopedia of Western Railroad History* (see bibliography) as having 12 miles of track. But the firm is also listed as narrow gauge, and the track abandoned in 1923 just at the time the United Railways (standard 56½-inch gauge) built to Keasey. If the firm operated after 1923 near Keasey and if so how it hauled its logs, was not within the parameters for this book. Oregon-American Lumber Company was a predecessor to Long-Bell Lumber Company with primary operations in Clark County, Washington. Its dates of operation, trackage out of Keasey into the forest, etc. is not stated other than it was standard gauge (56 ½-inch) track. The implication is that Oregon-American was no longer operating in the 1920's. The *Oregon Gazetteer 1931-1932* (R. L. Polk & Co.) shows that Oregon-American Lumber Company maintained a rented post office box at Keasey in 1931. Other data was not within the parameters for this book.

Both Inman-Poulsen and Oregon-American were operating in the Keasey vicinity according to many mentions of them in the lengthy interviews with Anton Lausmann in 1975-76. Lausmann had direct contact with the area until the Great Depression forced the end of East Side Logging Company in 1929.

Willamette Iron and Steel Works locomotive factory No. 14. East Side Logging
Company bought it new on April 12. 1924. Ran it regularly then sold it in 1928.

years but went for the Willamette brand as the
factory was very close therefore service would be
faster. The local firm built only 33 locomotives
between 1922 and 1929. A pair of them were two-
truck 50-ton models. Anton bought the first one. It
was factory Number 25. It became East Side Logging Company No. 107. He accepted delivery on
it on September 12, 1926.

Earlier, he bought a 75-ton Willamette. This
was in March 1924. In pictures, this locomotive is
seen as No. 102.

These Shay and Shay-type locomotives were
short wheel-base and were called "side-winders"
as they were gear driven by a shaft along one side
of the locomotive. The shaft applied power to
every axle including the wheels under the tender.
This meant that every wheel that touched the rails
was a drive wheel. On a 6 percent grade (which
was twice as steep as the toughest hill on the
entire Southern Pacific line), the Shay could pull
up to 63 percent more train than a conventional

rod locomotive. In addition, all of the driving
mechanism on the Shay was on the outside therefore it could be serviced or repaired easier than
some other brands of geared locomotives.

Anton also bought a 75-ton Willamette, factory No. 29, receiving it on April 22, 1928. He
numbered it No. 55. In July 1929, he took delivery on a used 45-ton Shay from one of his
second-hand equipment friends. Over the years of
operations, between his East Side Logging Company and his nearby Rock Creek Logging Company, he operated as many as nine locomotives
but not all at the same time.

Bob Lausmann designed the 75-ton engine
especially for East Side Logging Company. He
modified the plans for a 75-ton frame with weight
distributed over three trucks. This would support
larger capacity for fuel oil and the water storage
was increased to 4,000 gallons compared with
1,800 for No. 25. He equipped this with heavier
and larger gears and cylinders. He took his plans

LOCOMOTIVES OWNED BY ANTON LAUSMANN

MAKE	FACTORY NO.	ROAD NO.	WORKING WEIGHT	STATUS	DATE	WHERE USED
CLIMAX	1009	8	37 TONS	USED	CA1921	SUNSET/ROCK CR
CLIMAX	(SPECIFICATIONS UNKNOWN)			USED	CA1921	MCCALL/SUNSET
LIMA SHAY	1627	?	37 TONS	USED	CA1923	EAST SIDE LOG CO
LIMA SHAY	1732	?	28 TONS	USED	?	EAST SIDE LOG CO
LIMA SHAY	1902	?	45 TONS	USED	7/1929	EAST SIDE LOG CO
LIMA SHAY	2997	8	42 TONS	USED	CA1919	MCCALL/SUNSET
WILLAMETTE	14	102	75 TONS	NEW	3/1924	EAST SIDE LOG CO
WILLAMETTE	25	107	50 TONS	NEW	9/12/1926	EAST SIDE LOG CO
WILLAMETTE	29	55	90 TONS*	NEW	4/28/1928	EAST SIDE LOG CO

* MANUFACTURED TO LAUSMANN SPECIFICATIONS

Loco No. 102 burned oil instead of cord wood, puffs backward up a steep mountain grade with a long load of logs on disconnected trucks. Note the always coiled and ready fire hose on the tender.

to the Willamette works and discussed it with them where officials were enthused with his specifications so they built one for East Side Logging Company. (This was construction No. 29.)

Willamette Iron and Steel Works added Bob Lausmann's design to their catalog and built two more which they sold to others.

As was true for railroads everywhere, when a siding was built it was given a name. Because a siding was needed between Vernonia and Keasey to serve the Rock Creek operations, the United Railways named the siding "Lausmann Station." Switched to this by-pass track, Lausmann put down three miles of private track for the Rock Creek works.

Rock Creek Logging Company was operated as an entirely separate venture from East Side Logging Company but the two firms were geographically close together. The accounting, purchasing and other administrative affairs were all under Tony Lausmann's thumb from the single Portland office.

The first locomotive Tony owned was a 37-ton low pressure Climax. He bought it used about 1921. This was also a geared locomotive but of a different design from the Shay. (See the chapter "Common Woods Terms" in the front of this book.)

The Climax locomotive was first used with the Miller-Cox outfit and when Lausmann renamed that venture, he had SUNSET LOGGING COMPANY lettered on the tender. After he opened his operations near Keasey in Columbia County, the Climax went with him for use on the Rock Creek venture. To the confusion of historians, he never painted out the SUNSET name.

(CENTER) **Extra long logs on Rock Creek Logging Company disconnected truck train at Burlington log dump near Linnton on the Willamette River.**(LOWER) **Men take a break for the benefit of a photographer.**

The disconnected truck allowed great efficiency at low cost, compared to buying flat cars, for hauling logs on the private rail lines of the timber outfits.

Generally, logs had to be cut with two basic considerations in mind. If a customer required a special length then this had to be met. The other was attention to the length of the railroad cars on which the logs were to be hauled. If only railroad company cars were to be used, the logs couldn't generally exceed forty feet. If the customer required longer logs, these would have to be mounted on two cars at extra cost.

To get the most out of the private railroad, it was decided to use disconnected trucks. These trucks were merely dollies. With the dollies, logs of any length could be hauled. Anton bought

eighty sets. This was the equivalent of 160 cars except that one pair of trucks could handle a load of 80-foot logs which was double the length of a commercial car.

When the surveyors were planning the track layouts, the first thought was to use regular railroad cars. While this would eliminate the need to buy disconnected trucks, it meant that curves had to conform to public railroad specifications – the

Hayrack boom on spar tree used for loading logs onto trains of disconnected trucks at East Side Logging Company. Note two donkey engines at left.

57

Divide Camp of East Side Logging Company was like a small town. Bunk houses (left, rear); Storeroom (center, rear); Railroad field service shed (foreground); Cookhouse with dining room (right foreground); Storeroom (behind); Tony Lausmann private cottage (right, rear). See the camp from another angle on page 49.

curves not generally exceeding 15 degrees. A mainline locomotive could stay on the rails at 22 degrees and so could the cars. But with even one more degree of tightness, the whole train would jump the track. The geared locomotives, invented for the tight curves of logging railroads, could handle 30 degrees of curvature. To operate at a safety margin, the track was limited, with a very few exceptions, to 25 degrees for the curves.

Unlike standard cars, there were no air brakes on disconnected trucks. All the dolly did was roll with the train. But every truck had its own hand brake.

With disconnected trucks, the operator had to hire more brakemen. The engineers would stop the train at the top of a long grade to allow the brakemen to set all the hand brakes. On a so-called 40-car train – 80 disconnected trucks – this was not a quick job. The engineer, working closely with the brakemen, would run the train at a dead-slow speed while the brakemen ran along side keeping up with the moving train setting the brakes as the dollies rolled slowly by. The brakemen used their hands to turn the wheel that set the brakes just near the point of grabbing. Some hand brakes were stiff and could not be set by hand so a "hickey," a lever designed for the purpose, was inserted into the brake wheel to force the turn of the wheel.

Speed was very dangerous to a log train so

(TOP) Lima Shay No. 2997, Road No. 8 was a used locomotive bought about 1919 by Lausmann then lettered for Sunset Logging Company which was one of his operations before his East Side Logging Company days. (LOWER) The men loved to ride a caboose, of which the company had only a couple, being primarily for use on the main-line, here seen on the way down the mountain behind a train of loaded disconnected trucks.

trains were moved slowly.

Lausmann:

A train of loaded disconnected trucks was very fragile. If the engineer jerked with his train, he would pull the train apart. Our grades were often five percent and more. The grade to the top of the divide was nearly one mile long and it had some tight curves, some on trestles. As a train started down a grade, the brakemen either riding on the load or running along the side of the train, inserted his hickey to tighten the brakes.

The engineers treated these loads with extreme care for if they rushed their train faster than the brakemen could make final settings, all that load of logs could come down on top of them.

East Side Logging Company had only one serious runaway train and this was caused by lack of cooperation on the part of the engineer. In that instance, the engineer did not signal the brakeman in time and the 50-ton Willamette No. 107, roared down the steep divide grade and piled up in a heap in a canyon. No one killed. No one injured. But the locomotive was so badly wrecked, it had to be sent to the factory to be repaired.* There was no product damage as the train was merely a string of empty dollies – some on each end of the loco-motive – being moved getting ready to make up a train. Although the engineer applied full air brake pressure, the weight of the trucks roaring down hill pulled the engine along because, of course, disconnected trucks didn't have any air brakes.

Lausmann:

That engineer was so embarrassed that he had lost his train that he just walked away – never drew his pay. The locomotive had to be picked up, taken apart, loaded on rented flat cars and sent back to the factory.

Logging operators had to calculate into their operational plans such contingencies as wrecks and other equipment damage as well as time loss. The men idled had to be either paid or laid off. If there was a way to keep the crew together, as a good crew was hard to build, the men were kept even if it meant several days of no production. But

The 50-ton Willamette No. 25, Road No. 107 lost her footing on a down-grade and wrecked.

* While at the factory, this 50-ton locomotive was sold. Later, on another line, it was wrecked again, rebuilt a second time and lived out its days in Timber, Oregon where it was finally scrapped in 1950.

No. 107 was so badly damaged in the crackup it had to be returned to the factory for rebuilding.

if the logging operation was fairly small, and there was a major disaster, it could mean the bankruptcy of the company. East Side Logging Company was large enough to absorb the costs of the train wreck but it was costly in wages and room and board for those men who were pulled out of the woods until the wreck could be cleared.

Although the company owned some flat cars, most were not suitable to be used on the main line out of Keasey because they had been banged up in the woods.

Railroad equipment inspectors always had a field day when they showed up at the Keasey interchange. Armed with a pocket full of bright red "condemned" tags, they looked at everything – flat cars, box cars, cabooses – and seemed to love to hang their tags on the handle of the hand brake where it could not be missed.

Disconnected trucks being run on a main-line railroad was a taboo, except for the Lausmann operations of Rock Creek Logging Company and East Side Logging Company. There was a special

permit that allowed these log trains made up of dollies to run on the main line from the Keasey interchange to the Willamette River log dump at Rafton. (Rafton, the name of a railroad siding, does not appear on current popular maps.) If logs had been sold on a special order for another delivery point, then those logs had to be transferred from the disconnected trucks to main line flat cars at Keasey.

Lausmann:

We received a custom order for poles that would be used in the foundation for a large building on the Chicago River. We selected an area in the forest where we could get matching trees, all Douglas fir. We felled the trees then hauled the logs, which were 120 feet long, to the interchange on disconnected trucks. In the switchyard, the railroad crew and my men transferred these to main line flatcars. Three 40-foot cars were required for each load with the center car an "idler," merely taking up the space, to handle the overlapping.

With a private logging railroad the size of this operation, the company maintained a rail equipment repair facility at the main camp. On the payroll were men qualified to do all the servicing of locomotives, cars, disconnected trucks and the track itself. The Standing Order was to keep everything in trim for use on the private line – not

Extra long logs about to leave for the east for a special order, have been loaded on three flatcars for the main-line. The center car was an "idler" to take up the space. The logs were suspended over it.

maintain it to the national requirements for main line use.

If the private equipment did not meet the scrutiny of the main line inspectors, the logging company management was not overly concerned. This was never a big deal because the loggers owned very little in the way of cars that ever saw the main line.

East Side Logging Company owned a couple of cabooses, a "moving car," a pair of gondolas for hauling ballast, some oil tank cars and a few freight cars.

When oil was needed, a side-winder pulled a tank car out of the mountains to the interchange and left a pickup order with the United Railways station agent. The next Portland train hooked on to the empty tanker and dropped it at the bulk oil plant on the outskirts of Portland. Later, when full, the oil car was hauled back to Keasey where it was spotted on the logging company spur awaiting the next geared locomotive that could conveniently pull it back up to its parking track in the

Buy new or used? The longevity of logging equipment is well established therefore Tony Lausmann made the most of it by buying used goods – everything from old boilers, beer vats and locomotives to wire rope. (TOP) **A used boiler.** (LOWER) **Wire rope, also called cable**.

forest. The full car might stay at the interchange for days waiting on that key word: "convenient." United Railways could not charge demurrage (compensation) on privately owned cars left standing on privately owned track.

Making deals with his junk men friends and pawing through their yards was always a happy excursion for Anton Lausmann. On one trip, he bought a used 100,000 gallon Anheuser-Busch beer vat. This "tank" was installed on a hillside at his main camp, then his men laid track up another hill, behind the tank. A locomotive could push a full oil car up to the beer vat where the car could be parked. The oil was gravity drained through a hose into this king-size beer can. It always took several days to drain the tank car but this did not interfere with passing trains because the main

track passed this oil dump spur.

East Side's oil cars also came to the company by way of the junk dealers. The railroads dumped them as they only held 8,000 gallons which was uneconomical for transporting oil on the main line. They were replaced with the new 20,000 gallon tank cars that had become available.

Lausmann recalled:

These small tankers were in perfect shape and met the standards of the railway inspectors. I told my junk men to buy a few for me and they did.

A "moving car" was a large, heavy duty steel, specially built 60-foot flat car. It had 200,000 pounds capacity. It had air brakes so it was acceptable for main line use. East Side Logging Company needed such a car so Lausmann had one built by Pacific Car and Foundry Company in

(LEFT) **Track-walker made continuous rounds at East Side Logging Company to inspect track and to report damage or week spots.** (RIGHT) **In non-critical areas as on parking spurs, scrap track of various sizes were commonly used. Gaps between rails were frequent.**

1923. Some of these cars had double-trucks to support extra heavy loads. They also had fittings to which winch lines could be attached.

Also acquired were a specially built caboose and some box cars from M. F. Brady Company. The trucks under these were designed for logging railroads and would twist tighter than the 22 degree maximum for main line cars. He also had the trucks under the 2nd-hand oil tank cars adapted to his railroad in his shop.

Lausmann discussed the moving car:

> The moving car was important to us when we had to move an entire operation. A donkey engine would pull itself up on to the moving car on its skids with its winch. An entire donkey engine with skids was frequently between 60 and 70 feet long. If the load for the moving car exceeded its length, we added an "idler" car. On some trains we had to use two idlers, one on each end of the long load.

Many ideas were tried in the camps and on the logging railroads with the idea of saving time and money. One of these was to build a home-made miniature flat car which was just a platform mounted on a disconnected truck. The car, hooked to a gasoline powered speeder, was intended to haul up to four men and a little equipment for quick trips between the interchange and the various stops along the track.

Regrettably, the weight of the little platform car and its dolly was so great the speeder could not pull it. The little car was sometimes used on a regular train.

To solve the need for a quickie-trip vehicle, it was decided to buy a Dodge truck. The usual tires and wheels were discarded and flange wheels were installed. This vehicle was handy for the track patrols and for track repairs. The track walker took the Dodge, stopped at intervals along the track then dismounted and walked sections of the track looking for damage and weak spots, fallen snags, broken rails, washouts – anything that would be troublesome for a loaded train to pass over. If the track-walkers discovered some emergency, there were trackside telephones at intervals for him to call the main camp. The idea of using a regular truck on a railroad track was not new. In some locations, passenger cars with flange wheels ran regular schedules on main line railroads.*

---

* An example: The Pacific & Eastern Railroad running between Medford and Eagle Point used an Omnicar touring sedan that had been converted to flange wheels. It seated 7, including one passenger in the front seat. Luggage was in an open rack on the roof.

(TOP) **The "landing" for logs dropped at random around the spar tree.** (LOWER) **In the early days, cull logs called "conkies," were left where usually they fell or might be used as fuel for the donkeys if there was one nearby.**

A *Truck* Rides the Rails

The Dodge truck proved itself on occasions when important people visited the camp in rainy weather because the usual way to get one or two people from Keasey up the mountain was on a speeder. But the speeders didn't have roofs and were grossly unpleasant to ride in the rain. The Dodge had a roof. It was a solid vehicle but provided a rickety ride as it bounced over the uneven rails even at the track limit of only 15 miles per hour.

Lausmann:

A major guideline when laying out a railroad was to get good natural drainage so the track wouldn't cave in or wash away. We placed ballast whenever we could but the best ballast was very expensive. On one spur we used a lot of sand, which is excellent ballast but not as good as gravel. Along the Columbia River, where there was lots of loose gravel, we would scoop up a few loads to put on my railroad. These were the days you didn't have to have a *permit* every time you wanted to turn around.

The railroad was kept as simple as possible. It was a single track with switches only at loading areas or at a siding, and it was seldom that two trains ran in opposite directions on the same track. Even so, there was an informal dispatching system so it was always known where a train was.

Great care was taken to avoid collisions as a needless wreck might tie up the whole operation for days. Track clearance was usually obtained with just a telephone call to the offices along the line. When a train stopped at a field shed, one of the crew got off to call in for instructions – train orders. Every loaded train had several brakemen and the head brakeman was usually the "clerk" and did the telephoning as there were few conductors. The speeders also had to have track clearance.

There were no automatic or electric signals at the switches or at the top of a steep hill. But there were plain hand painted panels next to each hand operated switch.

At first, the logging operation was definitely a non-union organization. Few jobs were specialized and laborers in the woods did whatever job the crew boss ordered. The handful of specialized jobs included locomotive engineers, saw filers, machinists, blacksmiths and head cooks.

Many of the brakemen worked in the shops as helpers when they were not on a run. In most instances, the engineers did nothing other than operate the loco-motives and as Lausmann quipped, "read detective magazines and dime novels."

The firemen could very easily be a "grunt" (general helper) on days when the trains were not running. Some firemen were qualified to operate a locomotive but when they did this work, on rare occasions, they were still classed as firemen for payroll purposes.

On the wood-burning locomotives, the fireman hand-fed heavy slabs of wood into the fire box. It was a tedious job. He had to be careful, even with heavy gloves, as the slabs often became dislodged and shifted in the tender as the train creaked along the uneven rails. He had to keep his hands out of the way of falling fire wood. If his locomotive was wood fired, the engineer would run his locomotive onto a spur by a wood yard. It was the fireman's job to see that wood was cut and loaded into the tender. On occasion, general laborer's were detailed to move broken and "conky" logs from the woods to the wood yard and to assist with bucking them into proper lengths, usually 4 feet long and not over 10 inches wide. The wood was mostly fir which burned fairly fast and when properly vented, very hot. It left a lot of ash, which was the fireman's job to shovel out.

In the Lausmann camps, most engineers helped their fireman with the fuel loading. On the oil burning locomotives, the fireman's physical activity was considerably less and he too had time for reading dime novels.

Some camps had major repair shops. The machinery included lathes and other precision equipment such as shapers and milling machines. All the shops had power grinders. It was generally true that a single motor operated all the machinery which was set up in a straight line. Each machine was powered with a leather belt to the "line shaft" that ran just under the ceiling.

The main shop near Keasey was also equipped to put "tires" on locomotive wheels. These were the iron tires with the flanges which had to be shrunk on. When they cooled, the shrinking held them in place.

Nevertheless, derailing was frequent. Most were minor but some took many days to clear.

Lausmann, about his railroading operation:

Our most common factor that caused upsets was that we did not attempt to keep our railroad up to the high standards of the main lines. If there had to be a secondary factor, I'd suggest it was that our curves were too tight and engineers tried to take their trains around them too fast even through there were track [speed] limit signs posted. Our engineering could all go to pot if the operator did not follow instructions. In some places, if we said 4 miles an hour, we didn't mean 7 or 10. Sometimes the rails just spread apart due to various reasons. We'd get a crew together and rerail the cars. Often these derailings occurred in a remote section so blocking up a locomotive, or a string of loaded cars and getting them back on the track, was a time consuming and costly chore. But those fellows knew how to do it. I would hear about it only if it was severe and caused excessive down time or if there was a serious injury which was seldom.

My going out and looking at it was not necessary or practical because, normally, the boys would have the track fixed and the cars re-railed with the train moving again before I could get there.

My superintendents were responsible men but the best way to louse up a perfectly nice day was for a superintendent to telephone me in Portland and declare:

We have a train off the track!

My only answer would be: **Damn.**

We were in the mountains and we had some pretty tight and steep curves. We gave heart failure to the railroad companies if they learned that one of their cars got onto our track.

When private enterprise, like a logging company, built a rail line, a formal dedication was in order. When the work of constructing the line was completed for East Side Logging Company, there was a "golden spike" ceremony.

Lausmann:

Just about everyone we invited showed up. The President of the S. P. & S. loaned his private car. It was to bring Mr. John Churchill Ainsworth, President of the United States National Bank of Oregon, and several other bankers and timber company executives to the party. The ceremony would be at the Divide Camp at the top of the hill. At the last minute, the station agent at Keasey got a telegram from the trainmaster in Portland pleading:

Please don't let Mr. Davidson's private car
go up on that crooked Lausmann railroad.
The car is not adapted to take the curvature.

With this sudden change, the only way the gentlemen visitors could get to the party was for them to ride on the top of the water car. This car was merely an old flat car with a large watertight box mounted on top of it. Some planks were hastily placed over the top of the open vat for the gentlemen to sit on. A puffing little side-winder locomotive hauled this emergency passenger carrier down to the interchange then, with a ladder leaning against the tank, everyone climbed to the top. The men thought the method of transporting them was quaint but they had a lark. Although there was a safety rail around the end of the car, this railing didn't extend as high as the visitors would sit. This flat car belonged to United Railways but it was on more-or-less permanent loan to Anton Lausmann as the railroad really did not have any particular use for it. This car's trucks had been adapted in the East Side locomotive shop to take the sharp curves.

With the guests aboard, the locomotive was coupled to the car and off they all went up the mountain on the twisting track. The visiting financiers, with Anton Lausmann also on board, had several thrills on that outing which included holding their hats to their heads when they suddenly found themselves looking straight down over the edge of a high trestle.

*   *   *

To get across a valley, it was tactical to trestle the valley rather than to go around it. The trestles were built with a steam pile driver and most of the piles were taken from the forest along both sides of the right-of-way. The poles were of hemlock and small firs between 14 and 18 inches diameter.

Trestles on all Lausmann operations were built to main-line specifications. In the twelve years he ran railroad logging operations, he never had a trestle fail. (TOP) A trestle on the Rock Creek Logging Company operation. (LOWER) One of several trestles on the East Side operation.

Heisler locomotive on exhibit in Lewiston, Idaho.

The caps on the piles were all logs hewn in the woods close to the job. Sometimes the stringers were cut in a Lausmann mill or, if expedient, they were purchased cut to order. The trestles were built to main line rules on all Lausmann railroads over the years.

In some of the operations, it was possible to pipe water from a spring and to run it gravity-flow to the cook house or to a donkey position in the forest. If there were no springs available, then it was considered lower cost to haul water, 8,000 gallons at a time, in one of the railroad tank cars, than to sink a well and have undependable flow. It was common to run a water line across a trestle to avoid having to insert a pump in the line to get water back up the other side of a valley.

Appreciating the expense of building a trestle and the great damage that could result if a wreck occurred on one, extreme care was given the construction. Of all the derailings that occurred to logging trains, Tony Lausmann never had a loss on a trestle or had a trestle collapse.

Lausmann:

I think that in every instance of railroad wrecks and car

problems – my company's or anybody else's – it was a combination of trackage – mainly degrees of curvature – defective rolling stock and the human factor. No curve should exceed 25 degrees. But sometimes we had to go higher because of rugged terrain. New equipment could be purchased to fit any trackage but we were usually satisfied with second hand cars that did not get much modification.

The sidewinder Shay [Lima] and Shay-type [Willamette] were ideal for sometimes roughly laid track. There were alignment problems, ballast problems, washouts and sharp turns. Sometimes one or more wheels of a locomotive would be off the track, that is, air between the wheel and the track because of the unevenness of the rails. A rod-engine would lose its pulling power in such instances, whereas the Shays – *Oh*! *How I loved those Shays*! – would pull the load steadily because all the wheels were drivers. The use of the little old stubby-nose, cowcatcherless side-winder Shays guaranteed a successful log train operation – everything else being equal.

At the peak of Lausmann railroading, he had as many as nine locomotives but not all of them at one time. These locomotives were mostly Lima Shays and Willamettes. During the period, he would trade these engines just as easily as one trades in a used car today. <>

Owen-Oregon No. 2 pushes a load of logs on flatcars near Butte Falls.

One of many de-railings. The cleanup was always costly in time-loss because the railroad was often out-of-service for days. Logs were seldom damaged but nearly always some cars had to be rebuilt or replaced. If the total repair was estimated to be excessively lengthy, logging trucks were brought in to make the hauls.

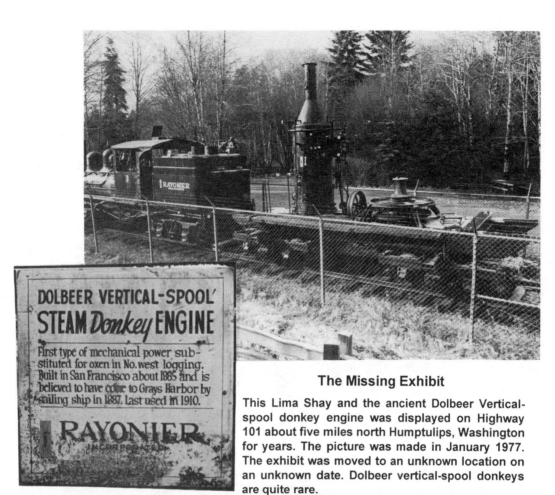

DOLBEER VERTICAL-SPOOL'
STEAM *Donkey* ENGINE

First type of mechanical power sub-
stituted for oxen in No.west logging.
Built in San Francisco about 1885 and is
believed to have come to Grays Harbor by
sailing ship in 1887. Last used in 1910.

RAYONIER
INCORPORATED

The Missing Exhibit

This Lima Shay and the ancient Dolbeer Vertical-
spool donkey engine was displayed on Highway
101 about five miles north Humptulips, Washington
for years. The picture was made in January 1977.
The exhibit was moved to an unknown location on
an unknown date. Dolbeer vertical-spool donkeys
are quite rare.

20 TON LIMA
SHAY LOCOMOTIVE

Built by Lima Locomotive Works...
Lima Ohio. 1910. Was last working
in regular use here in 1940.

RAYONIER
INCORPORATED

The Famous Donkies

From simple to complex models, the donkey engine was the King of the Forest for decades. This homely steam-belching engine supplied the power that was essential for operations.

—Photographs from the collection of John T. Labbe.

A donkey engine came without any roof but this addition, severely needed in the often very rainy Pacific Northwest, was later individually designed and built in the woods. The roofs were usually of second-hand sheet-iron.

(ABOVE) **Donkey at Simpson Logging Company, Shelton, Wash.**

(ABOVE) **Donkey at Benson Timber Company at Clatskanie, Oregon.**
(RIGHT) **Donkey at Noyce-Holland Logging Company, Kerry, Oregon.**

One-log load was photographed in July 1954 by the authors near Acme, Washington.

Loads of Logs

Loaded log trucks move nearly every day of the year from the forests to the mills. In the old days, high-profile loads, as top-left, frequently knocked the ends out of covered bridges. Pictures were made in spring 1990 at entrance to Boise-Cascade log yard in Medford, Oregon.

It takes a lot of logs to keep a big mill operating. (TOP) Shown is one mere corner of the log yard at Boise-Cascade in Medford, Oregon. (LOWER) It was economically difficult in the early days to haul logs from very small plots of timber. In one case, shown on page 104, two tractors needed nearly a full day to pry and shove some of big logs onto a flatbed truck. Today, some independent loggers and truckers equip their rigs with self-loading derricks. Photographed near a Boise-Cascade cold deck at White City, Oregon in February 1996.

MEDCO

Medford Corporation (MEDCO) had a number of predecessors as the various earlier outfits changed hands and names. Just after the First World War, when new businesses were forming, it was noted that the fruit growing industry in Oregon's Rogue River Valley was consuming upwards of seven million feet of lumber just to build packing boxes. The lumber was coming from a mill in the Klamath Falls area. But within sixty miles of Medford, in the Butte Falls region, there were large stands of pine from which the boxes could be built.

James N. Brownlee, a timber man formerly of Mississippi, and Millard D. Olds, who had been in the timber business in Michigan, went into business as the Brownlee-Olds Lumber Company, in 1922, in Medford. Each had been independently venturing into the timber business in the Medford area for a couple of years.

Brownlee had obtained a 33 acre site on the north side of Medford to put up a mill. He started building it in 1920. He also bought the Pacific and Eastern Railroad that ran between Medford and Butte Falls. Its right-of-way contained nearly six million feet of timber. By relaying some track to the mill site, there would be no difficulty moving logs from the forest to the mill.

Olds started cutting trees in a limited manner in the Butte Falls area then shipping his logs into the Medford mill on the railroad. The railroad, before the change in ownership, had operated a limited passenger and freight service but this was discontinued as the railroad now became a privately owned log carrier.

Jim Brownlee, who was in his sixties, sold his interests to Olds in 1923 then he retired to Mississippi.

James S. Owen, from Wisconsin, became interested in the Olds operation and after making a close study, proposed to buy it – railroad and all. In 1924 Owen incorporated as the Owen-Oregon Lumber Company. The railroad was split off as a separate entity and was named the Medford Logging Railroad Company. It was wholly owned by the timber firm.

Diesel-powered Osgood crane loader in the woods near Butte Falls.

Log pond and mill of Owen-Oregon Lumber Company in early 1920's. The mill, under this name, later Medford Corporation (MEDCO), provided steady year-around employment for area until 1989 when closed due to lack of logs. The site is presently a vacant lot.

Owen-Oregon started harvesting timber in 1924 using crews it housed at its camps in the forest. Of interest is that Owen-Oregon sought married men with families as opposed to single carefree itinerant bachelors. The company provided some family housing.

The firm did its best to stay up with the times therefore teams of horses gave way to tractors and lines run by donkey engines for yarding logs.

In short order, about twenty miles of track was put down to the landings in the woods. The old rod (Porter) locomotive that had been acquired with the line, pulled the 40 flat cars loaded with logs to the mill. As operations grew, more equipment was added. Forty additional cars were purchased along with a Baldwin locomotive. This rod engine would pull the log trains over the so-called flat lands between Butte Falls and Medford. The smaller Porter was used in the forest for the pulls from the landing to the engine-change point near Derby where the Baldwin took over. In the woods were two Willamette Pacific Coast Shay geared locomotives. A Lima Shay was kept busy just moving rails and ballast in the ever-expanding system.

In Medford, it was realized that the 33-acre mill site was being quickly outgrown. When word got out that the firm was exploring out-of-town sites for a larger mill, a committee of the town's important personages induced Owens-Oregon to stay where they were by making available sixty more acres that abutted the mill on the north and east – former county fair grounds.

Because a mill needs a lot of water – it had a ten acre pond in addition to other requirements – the city promised a special low water rate. In return, Owens-Oregon presented its water rights on Big Butte Springs to Medford, along with an easement through its property for a pipe line to feed the city's new water system. (Over the years the city, a little at a time, renounced its special low rate policy thus the mill had to pay the going commercial rate.)

There were numerous expansion projects, the largest requiring the sale of bonds to raise over $3 million for a new mill and equipment. A lot of the equipment was steam-operated from the power plant on the site. The boilers were fed by sawdust and scrap from the mill.

(TOP) The mill and log ponds of Owen-Oregon Lumber Company later Medford Corporation, occupied this real estate at the north end of Medford for over 70 years. After the facility closed in 1989, buildings were dismantled, hauled away and ponds filled. The vacant lot, photographed in February 1996, awaits development. (LOWER) Medite Corporation, the outgrowth of Medford Corporation, makes fiberboard in the plant built in 1973 at one end of the property.

MEDCO Willamette Geared Locomotives

(TOP) Construction No. 6, Road No. 2 purchased used date unknown probably in 1924. Scrapped August 1959.

(CENTER AND LOWER) Construction No. 21, Road No. 7 purchased used date unknown but after 1949. Sold June 1965 – exhibited in Railroad Park, Dunsmuir, California.

—John T. Labbe collection

(TOP) Logs have come in on trucks from the woods to landing near Butte Falls where they are reloaded onto flat cars (hidden behind loaded log truck) for train ride to the mill in Medford. (CENTER) Logs in the pond are sorted by size then moved into position to be milled. (LOWER) Lumber carriers haul finished product about the yard for storage or to ramps near the track to be loaded on freight cars.

The steam also supplied two 1,000-kilowatt turbo-generators for power throughout the plant. (It was said that the power generation was sufficient to light a city of 8,000 people.)

An innovative narrow gauge "railroad" system, on which two battery-operated locomotives operated, moved lumber around the plant.

For a general office, Owens-Oregon constructed an imposing neo-Georgian structure facing the street – Highway 99.

When the new facilities were completed in the spring of 1927, a grand reception and open-house was heralded in the local paper, the *Mail Tribune*. A bottle of champagne was broken over the first log, a fir, to enter the slip on its way toward the head-rig. The first product from that log was reported to have been a 2 x 8 x 16 board. It was estimated that the Owen-Oregon timber holdings would feed the mill for fifty years.

All was not well in the woods. Much of the old-growth pine used to produce box wood, was causing some buyers to complain of it having a poor appearance and texture. These complaints, true or not, were causing lost sales. The dimension lumber business was nationally off by 1927 and the lower sales from both divisions were reflected in the Medford plant's receipts.

The Great Depression was taking its toll. The two shifts a day on a 6-day week at the mill were reduced to only two and three days per week with only one shift. The pond was over-full and unsold finished lumber was stacked throughout the property. In time, the camps were closed and all the loggers were terminated except for a few who worked on special contracts and these were family men who lived at home.

In August of 1932, unable to meet its obligation, the Owens-Oregon Lumber Company passed from the scene. Many of its bond-holders formed the Medford Corporation (MEDCO).

Of importance to the firm was that by the late 1930's, among other matters, a serious look was taken at the earlier complaints about the quality of the box wood. One of their competitors wrote:

...despite the talk about the texture of ponderosa pine from Medford, he was unable to see any difference and would not hesitate to mix it with the produce of the Edward Hines plant [mill near Burns, Oregon]. ... These stories about our lumber being at a disadvantage because of texture or hardness are bunk.

Mill Waste Becomes School Fuel

In the late 1930's, a deal was made with the Medford School District to supply Medco waste mill ends and sawdust for fuel to heat the schools. The arrangement lasted about thirty years.

Medco was in a position to handle the increased demand for wood products with the opening of World War II. Of particular note is the fact that the U. S. Army decided to build a major infantry training facility in the desert about 9 miles north of Medford in the Agate Desert. The land, which was largely scrub growth infested with grasshoppers, rabbits and rattlesnakes, became Camp White. The Corps of Engineers advised Medco that it needed the railroad between the Southern Pacific interchange and the camp site for hauling supplies and later troops. As soon as the

Cold deck of logs. Some log yards are equipped with sprinkler systems to keep logs "fresh" during hot, dry, summer months.

"requisitioning" of the railroad was completed, there was competition for use of the track. A deal was made for the Army to use the trackage at night and Medco would use it for moving its log trains during the day.

In the postwar period, although Medco continued to use its railroad as its prime log hauling method, log trucks were growing in popularity. Medco had some trucks to use above its rail head at Derby. During the war, storage of cut logs became a concern so a sixty acre pond was dug in the Butte Falls area for temporary log storage.

After the war, the wooden fruit shipping box business came to a halt when the fruit growers and packers switched to cardboard boxes. While the dimension lumber business continued to be good, a new face appeared in the Rogue Valley that proved to be a sign-of-the-future: veneer mills and plywood layup plants.

In the woods, chain saws were replacing axes and hand saws. The donkey engines disappeared in favor of mobile trucks with yarding equipment.

The railroad laid more track and, in 1952, bought a new diesel Baldwin locomotive. The geared locomotives continued their chores on the rickety temporary tracks in the woods. But moving logs by rail was doomed because of the excessive expense.

Medco decided to build a cold deck facility for holding its logs on a site on the Agate Desert. It was called "Desert Pond" although it contained no water.

Finally, in 1962, the last train of logs – 19 cars – on the old railroad were pulled to the mill in Medford with the old Baldwin rod engine. Log movement after that time was by truck, the last

few miles using the old railroad right-of-way from which the track had been salvaged.

The Lima Shay locomotive was sold for scrap. The two Willamette Pacific Coast Shay locomotives went into museums. One is still seen in the Railroad Park just south of Dunsmuir, California. The other is now in the Medford Railroad Park but for years, this locomotive was displayed in the city's Jackson Park. In 1986 it was moved to the city's Railroad Park for its permanent home.

The old Baldwin steam rod engine went to Willits, California to operate on the California and Western Railroad commonly called "The Skunk." At last report, it was still making runs to between Willits and Fort Bragg.

In 1973 a particle board plant was built on the property north of the original mill.

In a relatively short period of time, several things happened. In 1984 Medco was purchased by Harold Simmons of Texas and an announcement was made that the company would log only its own holdings. The mill ceased all operations 1989. The idle plant was dismantled and much of its equipment sold. At the present time the cleared land stands in alfalfa.

A fiberboard plant built in 1973, on one end of the property, was renamed Medite Corporation and continues to run. The veneer mill near the city of Rogue River continues to operate.

For an aerial photographs of Medite's Medford fiberboard plant, see page 78 and for the Rogue River mill, please turn to page 141.◇

Remembering the Sidewinder

—John T. Labbe collection

The locomotive got the name "sidewinder" because all of the power to its wheels was applied from a drive shaft and gears on the outside along the right side of the engine. While conventional "rod" locomotives of equal length might have 4, maybe 6 drive wheels, this gear-driven short wheelbase logging engine had power to all 12 of the 36-inch wheels. While usual locomotives had wheels with flanges four inches wide, the sidewinders had 5½-inch flanges all the better to grip the rails.

When the Southern Oregon Chapter of the National Railroad Historical Society met in Medford in late June 1986, a guest for the day was 81-year old Elvis King. Decades earlier, he had been a fireman on "Old No. 4," the Willamette Pacific Coast Shay "sidewinder" locomotive that had been bought by Owen-Oregon Lumber Company.

King and the engine arrived in Medford within a few months of each other. The company took delivery of the locomotive on February 13, 1925. It was brand new. King arrived on September 16. He was 20. He was hired to be a fireman.

On the way out of the woods, down the steep grade, King recalled looking back at the end of the 20-car train which was above him.

The engineer had to be real careful going down that grade – very slow – for if the train ever got going too fast you could never stop it.

Unlike many logging trains that used disconnected trucks – no brakes except on the locomotive – the Medco trains had air brakes on every flat car allowing for greater control when going down grades.

For his work, Elvis King started at 50¢ an

Medco's "4-spot" (TOP) moving logs near Butte falls, Oregon. (LOWER) The last day of operation before being abandoned, the historic locomotive hauled a few dignitaries on a sightseeing trip.

(TOP) MEDCO No. 4 ("the 4-spot") was exhibited in Jackson Park in Medford for many years. It was later moved (CENTER) to new Medford Railroad Park where it is part of a display along with a number of older railroad cars.

Medford Railroad Park is operations center for Southern Oregon Chapter of National Railway Historical Society. In addition to the static exhibits of full-size rail equipment, the park has over one mile of 7½-inch gauge track where hoards of visitors ride free on the miniature trains in summer months. So important to railfans is the park, that a book, *The Sewer City Short Line* was written about it. Before being a city park, the property was part of the City of Medford sewage disposal system.

hour with room and board. He lived in a bunk-house at one of the forest camps.

It was the incredible power of the sidewinder locomotive – 174,000 pounds working weight – that made King proud of his work. As a fireman, he was responsible for controlling the mixture of air and fuel oil that entered the fire box. He also kept track of water levels and pressure gauges.

(The engineer drove the train by working a lever – throttle – that determined the amount of steam that was admitted to the three driving cylinders and a lever that controlled the length of the piston stroke – short stroke for speed. At that time, engineers got a dollar an hour)

Some steam was piped back to the oil tank to warm the thick fuel so it could be pumped through a feed pipe into the fire box. It was a common saying that on cold days, the fuel might be as "thick of molasses" until it was warmed.

King's Willamette Shay locomotive was a "superheater." A superheated system had an extra set of tubes that took steam back through pipes in the fire box to reheat the steam. King recalled that superheated systems delivered more power but used less fuel. The gear-driven locomotive had the same power going in either direction.

Keeping a good load of water on steam locomotives was always a challenge for road engineers had to compute how often along the track a water tank had to be installed. The Willamette Shay type engines on the Owens-Oregon system held 3,000 gallons.

For trips heading back into the woods, the train's empty flat cars were herded ahead of the locomotive.

In the forest, logs were gathered and brought to the landing which was alongside the track, then the logs were loaded onto the cars. Once loaded, the sidewinder hauled the train to the interchange to await the rod engine. The system, in the early days, called for two 20-car trains to work at the same time. One train was in the woods loading or hauling logs to the interchange, while the second train was on the Medford run.

Stalking Raindeer

King recalled that during deer season, a number of loggers, armed with rifles, mounted the first car of the train as it went back into the woods. When deer were spotted, usually as the train rounded a twist in the track and the deer were surprised, some deer froze while others loped off through the underbrush. Shots would ring out in hope of scoring a hit. But the rocking train on the often rickety rails, moving maybe 6 miles an hour in the forest, provided an unsteady platform for shooters who did not often score any hits. When a deer was downed, a special flag was waved back to the engineer who would stop the train so the carcass could be hauled aboard.

There was a unique method of blasting soot out of the stack on the engine. King recalled that after he shut down the damper, he would throw a hand full of sand into the fire box. The vacuum created would suck the sand up through the flues knocking out the soot. This was called a poor-man's method of sand blasting.

King did not recall his engine ever having to go back to a shop for repairs. The usual daily oiling and greasing of the joints was done in the field between runs.

Medford Corporation donated its old sidewinder No. 4 to the City of Medford to be used as a public display. It can been viewed in the Medford Railroad Park, off Table Rock Road. Of the six Willamette Pacific Coast Shay engines extant, MEDCO 4-spot is the only one in Oregon. <>

Statistics

Seven of the 33 Willamette Pacific Coast Shay locomotives built had but single owners. One of them, Construction No.18, Road No. 4, started and ended its life under the control of Owen-Oregon Lumber Company, later Medford Corporation.

Fighting Forest Fires

Load 'em up, fly to target, "bombs away," head for home.

Pick up another load, take off, "bombs away" head for home. Do it again and again and again.

But these bombers are not carrying high explosive block-busters these days. Each trip out, each bomber delivers between 800 and 3,500 gallons of retardant, depending on the type of airplane, to a forest fire. *

World War II era combat planes were built for hauling big loads and after being converted to be flying fire engine tankers, they still carry big loads.

Ever since the end of the war, many precision bombers then destined for the scrap heaps, have been kept flying after conversion to fire fighting tankers.

From wartime long-range patrol missions over the South Pacific searching for Japanese, a converted Catalina flying boat ("P-boat") waddles

---

War-Time Bombers Still 'Bombing'

After World War II, the most popular big aircraft to be converted to forest fighters:

B-17	Flying Fortress
B-24	Liberator
C-54	Loadmaster
PBY5A	Catalina
P2V	Neptune

Smaller former military planes that have had fire retardant tanks added include:

F7F	Tiger Cat
B-25	Mitchell
B-26	Marauder

PBY5A "CATALINA" Navy Patrol Bomber

---

* Years ago, small planes were the fire fighting "bombers' and there are pilots still around who remember that the best load was only 150 gallons.

FIRE

The Silver Complex Fire in Southern Oregon in summer 1987 burned 96,240 acres. While the smoke nearly blotted out the sun, the ash was so thick it rained on Medford streets like snowflakes.
—Medford *Mail Tribune*

PB4Y2 former Navy long-range bomber, also known as a B-24-D in the Air Force, was converted to a forest fire-fighting bomber. This airplane was lost when it ran out of gas and was forced to land in a lake. —Gary Austin collection

Crew of forest fire-fighting World War-II B-17 heavy bomber, with suitcases, are ready for an overnight trip.

down the runway looking like a gravid goose. It lumbers into the air its belly loaded with fire retardant. The P-boat's slow speed allows it to cruise through forest canyons with ease, banking at near tree-top level, to splash or spray its chemicals on forest fires.

The "pregnant guppy" that zooms considerably faster is a former carrier-based Gruman terror of the skies, the F7F Tiger Cat. Designed for a 2,000 pound bomb hung amidships plus two 1,000 pounders under the wings, this sleek airplane now sprays at fires from its 800 gallon belly tank. Powered by twin 2,150 horsepower radial Pratt & Whitney's, the Tiger Cat was planned to give stunning gouges on the Japanese home islands had not the atomic bombs ended the war.

Another Navy plane but of Korean War vintage, a P2V Neptune patrol bomber, carries up to 3,000 gallons of "fire-mix" and is a powerful addition to this provisional battle squadron.

One of the pilots, a tall, gaunt man who learned to fly in 1942 with the U.S. Air Force, served as a B-17 instructor then flew B-29s over Japan. He declared that he was an arsonist in the great incendiary raid by more than 800 B-29s that burned most of Tokyo, but now he is a flying fireman.

Numbers of fire tanker pilots were former military pilots and some gave years to flying passengers for the airlines. After the war, the most popular passenger carriers were the famous twin-engine DC-3s and the four-engine DC-4s. As the 4s had the more power and the greater carrying capacity of the two, many of these planes had become weary of hauling people and were converted to retardant tankers. Likewise, as the years passed, some DC-7's had their seats removed and grew giant bellies for holding retardant. For several years, the former United Airlines deluxe plane *City of Los Angeles*, now with a belly tank, was based in Medford.

(Experiments were made with the DC-8 jet liner and a number were converted for fire bomber use. But after trials there were mixed opinions. Among the jet's critics was one claim that the jet could not be slowed sufficiently to be effective.)

These privately owned flying tankers are chartered by the U. S. Forest Service and fly from home bases throughout the country to airports equipped with fire-fighting facilities near fire zones, when needed. Flying for the Forest Service is an exciting, serious, dangerous business and is not undertaken lightly by the airmen. Pilots must use the same professional approach over a raging fire as was done on a wartime mission dropping blockbusters. The approach and pull-out is all important. Now and then one reads of a forest fire fighting bomber that crashed killing the crew. The pilot that thinks buzzing tree tops is a lark – the fly-by-the-seat-of-your-pants type, might splatter himself and his aircraft all over a hillside. Flying fire fighting bombers is a precision business. The old war-birds are ideal for low-level forest fire bombing.

Although designed for high latitude precision daylight bombing, four-engined Flying Fortresses also flew wartime missions at wave-top level slipping in and then slipping out on missions in the South Pacific. They also performed admirably at

A former passenger liner, this DC-7 was converted to hold 3,000 gallons of fire retardant.

tree-top level sneaking in under radar on special missions in Germany. In the Battle of the Aleutians, B-17's and PBY's became dive bombers over Kiska. It was the B-24 that flew the historical raid that wove through the tall chimneys in the Ploesti oil refineries of Romania. All this "track record" makes these groaning bombers suitable for fighting fires.

Forest Service flying regulations call for minimum altitude of 75 feet over the burning treetops.

The Costs of A Forest Fire

Thousands of acres of marketable timber can be lost in a single forest fire and nearly every year forest fires are started by summer storm lighting strikes or by careless persons. Some fires are purposely set by arsonists. One fire was set in a forest by the mother of a young firefighter because she knew her son needed the work. Fighting forest tires is costly and can be deadly dangerous when conditions on the ground suddenly change. Wind changes have cost many lives when some crews on the ground are caught and burned over by fires.

(ABOVE) DC-7 former air liner, now a fire-fighting bomber (Tail No. 62), is the airplane shown in color on the back cover of this book. (LOWER) DC-6 former passenger planes, converted to fire retardant carrying bombers, on flight line at Medford Fire Center at the Medford - Jackson County Airport ready for duty.

The Oregon State Forestry Department set up a 1½-mile long line around the Logtown Fire that had burned 270 acres. The fire moved through timber and brush rapidly as it was whipped by winds. There were about 100 men on the ground from State Forestry, Rogue River National Forest, Jackson County Parks and men from nearby industries.

Retardant tankers dropped 21,600 gallons on the fire and five bulldozers were working on the lines manned by personnel from Jackson County Road Department and Lininger and Tru-Mix Construction Companies.

A small helicopter was used to make water drops refilling from convenient ponds in two nearby creeks.

The fire crews included men from KOGAP Manufacturing Company (a plywood plant), Timber Products (mill) and McGrew logging as well as eighteen high school youths who had been trained by the Forest Service.

The fire started along a road that runs alongside the Logtown historic cemetery but there was no damage in the cemetery. Most of the land burned was Bureau of Land Management land. Al-

though there were four homes nearby, all escaped.

Seventeen loads of retardant at about 2,500 gallons for each drop, were credited for bringing a fire under control near the city of Talent in Jackson County. The blaze got started near mining equipment parked on a rural road but the equipment was not being operated at the time.

Working in high winds was the challenge of the firefighters as there were several homes along the road that for awhile were in danger.

At the peak of the fire, about 100 firefighters were on the scene. These were from State Forestry Department, Rogue River National Forest, Jackson County Fire District No. 5, Talent City Fire Department and Jackson County Parks. Because the fire was in an area of private residences, crews were on standby from other fire districts.

The fire was mostly in timber and heavy underbrush on steep terrain that made access by vehicles difficult. Aerial bombers were of prime importance in containing this fire.

At an airport's fire fighting base, are special tanks maintained to capacity by the Base Supervisor of the Forest Service. These tanks are filled with a liquid mix of concentrated ammonium-

Unique KAMOV helicopter was developed about 1965 and often used in agriculture spraying operations. As shown here, a KAMOV was adapted to pick up, carry, then dump water from a large bucket on forest fires. Experimenting was done under the name of "Command Helicopters" from the firm's headquarters in Central Point, Oregon. The work stopped when the pilot, while under contract to the Forest Service to fight a fire, was killed when he allegedly flew too low, clipped a tree and crashed. The blades of the KAMOV are counter-rotating thus no tail rotor.

sulphate fertilizer and water. The Medford base can create up to 125,000 gallons of retardant. The special concentrate is shipped in by truck from plants in California, Washington and Idaho.

Years ago, most forested lands were protected ("watched") by people who spent night and day in the lookout towers. By now, most of the towers have been dismantled as obsolete and too expensive to staff, in favor of patrolling aircraft and radar spotting of fires.

During some fires, a Forest Service aircraft

—continued on page 97

(TOP) Water tank on left and fire retardant tanks at Medford Fire Center. The Center is operated by the U.S. Department of Agriculture, Forest Service. (LOWER) Belly tank especially fitted to the fuselage of a DC-6.

All in A Day's Business

Fire fighting tanker aircraft from Medford took to the skies to drop retardant on the Logtown fire then turned to aid the Grants Pass district with the 40 acre Jones Creek fire. The fire was reported at 5:15 in the afternoon and was controlled by 8:30 the next morning. The tankers dropped 14,000 gallons of retardant on the fire.

Tankers were also dispatched to the Klamath National Forest where 23 fires were started following lightning strikes.

Ooops!

Diamond Lake, a disabled B-24 and Mt. Thielsen

While returning to the Medford fire fighting air center after dropping retardant on a fire in the Willamette Valley, a weary PB4Y2 (B-24-D), with its four engines sputtering, gave up its ghost. The pilots expertly missed Howlock Mountain (8,363 feet elevation) and Mt. Bailey (8,351 feet elevation) and Mt. Thielsen (9,182 feet elevation) and dropped their lame bird into the shallow end of Diamond Lake (5,183 feet elevation). This is about twenty miles from Crater Lake National Park. The remark of old-time barnstormers was surely in the pilots' favor: "Any landing you can walk away from is a good one." But in this case, they went ashore in a Sheriff's boat.

What happened:

The converted B-24 simply ran out of fuel and had to ditch in Diamond Lake. The pilots exited the plane easily as the plane's fiberglass nose broke off on impact. When they saw the nose floating nearby their first through was that they had hit a boat. Witnesses at a YMCA camp, in front of which the landing took place, called it a "beautiful job of landing." The plane filled with water nearly up to the level of the wing as the plane sat on the bottom of the lake. The B-24 is a normally low-slung airplane.

The evening before, the two pilots had finished their work and had landed at the Medford base after dark following a flight from Lakeview. The lights in the flight office were off and there was no one at the field so they went home. In the morning, they loaded out for the retardant drop in the Willamette Valley. On takeoff, the pilot remarked that the airplane seemed "spirited" and easy to fly. The B-24 had not been refueled thus the gross weight was light. It was when the plane ran out of gas over Diamond Lake, on the way home, when the pilots realized why their engines quit. A B-24 can hold up to 3,000 gallons but on the local fire calls, the fuel load is about 800 gallons. This is to put the fuel's weight into retardant capacity. The next morning they climbed aboard for another day's work never remembering their mutual omission. Purely stated, they ran out of gas and ditched the airplane. Not only did they lose their airplane, both men were fired.

Because the landing was in an environmentally sensitive lake, and the lake being a popular vacation spot, it was determined that the only way to get the plane out of the water was to winch it ashore then dismantle it. The parts were hauled back to Medford where most were junked. *

---

* This airplane was a later model B-24 than commonly seen as it had a single fin and rudder. Data about the PB4Y (B24) is from *Janes All the World's Aircraft* 1943-44 Ed. Some of the information about this mishap is from an interview with the co-pilot, Gary Austin, in Medford on February 19, 1996. The picture of the B-24 resting in Diamond Lake is from the Gary Austin collection.

Formerly a main line passenger plane, this DC-7 air tanker-bomber flies gracefully whether with passengers or with a load of fire retardant. The airplane is easy to control at low altitude shown here as its drops up to 3,000 gallons of fire retardant on the Haner Butte Fire in Deschutes County on June 8, 1992. The site is near Highway 97. This photograph appears in color on the back cover of this book.

—U. S. Forest Service

(THIS PAGE AND OPPOSITE PAGE) **What looks like a war-time *Kamikaze* pilot on his mission to glory, is really a DC-7 tanker on a "bombing run" west of Talent, Oregon in the Anderson Creek area on July 24, 1977. Medford *Mail Tribune's* photographer Glenn Cruikshank made these battle-action pictures from an Oregon State Forestry Department helicopter. The air-tanker was from the Medford Fire Center at the Medford/Jackson County Airport less than ten minutes flying time away.**

acts as spotter and guide-in-the-sky talking by radio to the bomber crews directing the planes to the most advantageous angle from which to drop their loads.

When a call comes that a retardant drop has been ordered, the ground crew begins to fill a tanker while a dispatcher is on the radio getting locations. The "alert" plane – the first one to get off the ground – is parked near the tanks of retardant so it can be loaded quickly. Water in one tank and retardant in another is mixed to formula by special measuring valves on the pump. A hose extends from the pump to the waiting airplane. The planes, already fueled, are ready for takeoff in about eight minutes of a call.

On reaching the drop zone, tanker pilots adjust air speed to about 120 knots (about 130 mph) and go for their working altitude of about 75 feet over the forest. This speed, depending on the type of aircraft, and altitude, allows the most effective spread when the retardant is dropped.

Sometimes the cargo is splattered directly into a fire while on other circumstances, such as "crowning" fires in the treetops, the drop is made ahead of the fire. The retardant, being water-based, can evaporate and become ineffective if dropped directly into a very hot area.

In very troubled summers, a single fire base might deliver as much as 1,000,000 gallons of retardant on fires within a single month. On one 24-hour day at the Medford base, there were 38 trips that dropped 114,000 gallons.

Flying firemen are on active duty from 8 in the morning until 8 at night seven days a week, but during a fire, the day can be extended to include all the daylight hours. Are retardant drops made during the night? The answer is that it depends on the fire and operating conditions.

The "Old Man" who is falling to a splash landing. This World War-II B-17 heavy bomber was converted to drop up to 3,000 gallons of fire retardant.
—U. S. Forest Service

Often, during a quiet summer, pilots complain, "the only thing wrong with this job is the hours of boredom. You might sit for 12 hours a day and wait for a fire to happen."

One of the pilots related how he had flown cargo planes for eight years for the Air Force then five years with passengers for Western Air Lines but he prefers aerial fire fighting. But he allowed that roaring into a fire and dropping loads of retardant into a raging inferno at very low altitude at high speed is not a particularly safe way to spend a summer afternoon. <>

Erickson Air-Crane Helitanker

In 1991, Erickson Air-Crane developed the most efficient and effective fire fighting system available.

The Erickson Air-Crane 2,000 gallon Helitanker combines the capacity of a winged tanker aircraft with the turning time and accuracy of a helicopter. A powerful snorkel system refills the tank in less than one minute from any nearby water source thereby eliminating the requirement for fixed wing tankers to fly back to the airport for another load.

From the cockpit, the pilot mixes the appropriate amout of retardant to add to the water as the Helitanker flies to the drop zone, then the pilot selects the rate and amount of the drop. The Helitanker can deliver as much as 30,000 gallons per hour with a result of lower cost per gallon than any other system. Helitanker picture courtesy of Erickson Air-crane

Goodyear-Nelson Hardwood Lumber Company

and Catastrophic Fire at Mill No. 2

Victor Nelson and Frank Goodyear, both in their 50s, of Sedro-Woolley, Washington had been operating their three mills with success for a number of years. The author "discovered" them after he took up residence and operated a commercial photography establishment right after the close of World War-II. These men had two mills in Sedro-Woolley, one on each side of town with both on the Northern Pacific Railroad. There Mill No. 1, on the west side, which handled mostly pine and fir up to about 12-inches diameter. Their No.2 mill took logs upwards from about 10-inches. The product from these mills went into the general lumber market.

Mill No. 3 was at Port Angeles on the Strait of Juan de Fuca. The Port Angeles mill cut birch, alder and maple, which grew on the Olympic Peninsula. The product from this mill was all consigned to the Angeles Hardwood Company in Los Angeles. This lumber was used in the furniture industry. All three mills included dry kilns.

About May of 1947, Frank Goodyear telephoned the author and asked if I would stop by for a chat to see if a series of photographs might be made of his operations to illustrate some sales brochures he sought to publish. A business arrangement was concluded and the camera work got started. This was the first time the author had been in any lumber mill therefore there was much to be learned. The photography started at Mill. 1, then jumped to Port Angeles on a time table that was agreeable as Goodyear had to be there to point out what was to be pictured.

Vic Nelson did much of the office work while Frank Goodyear might be called an "operation manager." Goodyear went from mill-to-mill, talked with his foremen and workers and held conferences with business visitors who came to the mills. While Vic was a quiet type, Frank was just the

**Sedro-Woolley's Goodyear-Nelson Mill No. 2 burned in June 1947.
The cameraman arrived about 6:15 a.m. This was what he saw.**

opposite.

Many dozens of pictures were made, all on a 4 x 5 inch Speed Graphic camera. As each lot was completed, I'd call for an appointment to have both Vic and Frank available to sit down and look at the results. The work progressed in this orderly manner.

In the immediate post-war period, this was a small town of about 2,500 people. It has always been a town one could get lost in easily. This was because the two original villages, Sedro and Woolley, on each side of State Street, had early day promoters who did not want competition from the other. Therefore, all the streets were offset – no through streets across State Street..

The town is about six miles northeast from Mount Vernon but closer to ten miles by road. The town was founded based on forest products – saw mills. The Sedro side of town was platted around 1884 and underwent several names, including

Volunteer fireman, most without hard hats on this morning, and helpless to fight the fire or salvage any equipment, could only stand around and wonder about their leaking hose. The department was well trained for the time, had expert chief, Jack Hebert, who had invented the Hebert Hose Clamp that was universally used.

"Bug," but the local folks didn't think much of it. Because there was a lot of red cedar in the area, and a number of shingle mills processing that wood, the name *Cedro* (Spanish for the word "cedar") was offered to the Post Office Department. For its own unique reasons, the department papers came through as "Sedro." This post office operated from December 7, 1885 until it was combined with Woolley, which operated "across the street" between May and December 1890,

The rival town started by Phillip A. Woolley, on the east side of what became State Street, thrived in intense competition with its neighbor. Finally, by official vote of the peoples of both bergs, in December of 1898 – believed to have been plausibly instigated by the post office – the

two names were hyphenated to form "Sedro-Wo-olley." This is the only place in the world with such a name.

To provide a typical "birds-eye view" of the largest of the three Goodyear-Nelson mills, the author chartered a small airplane and a pilot. The object was to shoot aerial pictures. It had been decided that the better picture would be of Mill No. 2 because it had been more carefully laid out. After the flight, we opened the photo-lab to process the film and make prints.

A little after six o'clock the next morning, the siren on the town's city hall started to wail. This was the signal for all the members of the volunteer fire department to rush to the fire engines (2) that were kept in the city hall's garage. The

A fellow who lived on the outskirts of town needed some extra money so he hired a gyppo logger to come by to look at his small stand of cedar trees. The trees, once felled, went to a local shake mill.

town had a 1938 Ford fire truck and a 1925 LaFrance. Crew No. 1, in the Ford, with the usual parade of indi-vidual cars carrying the volunteers, headed for the huge, black, steady plume of smoke that was coming from the northeast corner of town. The first thought was that the fire was at Skagit Steel and Iron Works, which squatted on a large acreage in that direction. The fire was in Vic Nelson's and Frank Goodyear's Mill No. 2.

The author and his wife, Margie, lived in an apartment over the Safeway Store in the middle of town. When the siren sounded, I looked through the window and announced that the smoke looked like a plume from an atomic bomb.

I got dressed in double-time, beat it to the car, which was regularly parked in the lot behind the store, then drove two blocks to the photo-lab for the camera. In another few minutes I arrived at the mill in my used 1940 Ford car. The fire was an exciting sight to the eyes of a free-lance camera-man and stringer-reporter for the Seattle *Post Intelligencer.* Me!

The mill was totally engulfed in flames. I had loaded a 12-shot film pack into the camera which made shooting faster and easier than using the usual 2-shot cut film holders. I made pictures from various angles, then went back into town, about a quarter-mile, to the photo-lab. Margie, had proce-eded me and had chemicals "up" (to temperature) ready for the film. While the film was in the DK60a developer, I telephoned the City Editor of the *P-I* with the first take on the story and received a clearance to cover the fire for the paper. This gave me first option for selling the story and pictures for it was necessary to protect this fire story from neighboring photographers who were sure to show up. After an hour or so, several did.

Sedro-Woolley was only 70 miles from Se-attle. But this was before Interstate-5 was even a dream. The only way to get between the cities was on old Highway 99. This road, two lanes, wound around the hills and through the middle of all the towns (town speeds from 20 mph to 35 mph) so the drive was always close to 3 hours. With fol-low-up camera work yet to do, we packaged our photographs, addressed the envelope to the *Post Intelligencer* and sent it to Seattle by bus express. The express charge was $1.03 including sales

Pictured Day Before Fire

NOW RUINS—This aerial photograph of the Goodyear-Nelson Hardwood Lumber Company's Plant No. 2 at Sedro Woolley was taken Monday, less than a day before the plant was destroyed by fire. Night watchman discovered the blaze on wall between boiler room and fuel house about 6 a. m. Fire still was burning eight hours later. (Story on Page 1.)

OUT OF WORK—These lumber company workers posed for group picture day before disastrous fire. Now two-thirds of them are temporarily out of work. Others are employed in company's No. 1 plant on otrer side of Sedro Woolley. No. 1 plant was hit by fire in December. Most of current fire loss is covered by insurance and company may rebuild. —(Photos by E. T. Webber, Sedro Woolley.)

tax. But the bus was even slower than taking the car for the package had to be transferred from the Mount Vernon & Anacortes Stage Lines at Mount Vernon to the next southbound North Coast Lines bus that eventually arrived in Seattle. (This was before Greyhound bought in.) By arrangement, the *P-I* would have a local courier meet the bus.

A picture of the mill in total involvement with fire appeared in the newspaper the next morning. Inside, the editor decided to use the aerial picture of the mill and a snapshot I had made during a lunch break, just one day earlier, when Frank Goodyear rounded up a large number of the workers and had them pose for a picture. The paper said:

Clipping from Seattle *Post-Intelligencer* the morning after the fire.

These lumber company workers posed for this group picture day before disastrous fire. Now, two-thirds of them are temporarily out of work. Others are employed in company's No. 1 plant on other side of Sedro-Woolley.

No. 1 plant was hit by fire in December.

—Photos by E. T. Webber, Sedro-Woolley.

A night watchman discovered the fire on a wall in the mill between the boiler room and fuel house about 6 a.m. He promptly notified the night duty police officer, who hit the switch to sound the siren for the volunteer firemen.

The fire (TOP) by 8 o'clock had substantially moved to the next buildings. (LOWER) By 10, there was little but frames left of the main building.

(TOP) **Leaking fire hose made for a muddy scene on an otherwise dry day. The only peaceful parts of the scene are the undisturbed dandelions.** (LOWER) **By noon, the fire was well into adjoining buildings and a few folks from town still stood around.**

Sedro-Woolley had two cops. The Chief, who was also the Skagit County Constable, and one regular patrolman. The patrolman mixed with the crowd (his neighbors) to keep order if anyone was out-of-order. There was no out-of-order.

In such a blaze, there is very little a small town volunteer fire department can do. The men ran some hoses bringing water from a hydrant a block away, then they watered down dry grasses and brush around the back side of the burning building to keep the fire from spreading to half-a-dozen houses about 200 feet away. Others of the volunteer company set up a pumping line to take water from the mill pond if it was needed.

Most of the fire loss was covered by insurance.

For this book, we reproduce two pictures from the original newspaper feature and include other pictures made fifty years ago that have never been previously published.<>

2 p.m. – Eight hours after it started, the fire was still active.

Removing Logs From the Woods

The usual method for taking logs to mills has been to carefully carve access roads through the forest following the most direct route to an established highway. These access roads are given Forest Service or BLM route numbers. Over these routes, log trucks remove the felled trees. Many of these roads deteriorate in years following the logging operation and become impassable except for hunters with 4-wheel drive high-center vehicles. The object is to allow most of these routes to return to natural state which they often do.

(TOP) **Cold deck of Goodyear-Nelson Lumber Company was at Mill No. 1, about two miles across the city from Mill No. 2. (LOWER) Two loaded log trucks have arrived at the pond of Mill No. 2. Buildings in background are part of Skagit Steel and Iron Works. The firm made logging equipment.**

(TOP) **Mill No. 2 with windows shuttered, is closed for a winter nap.** (LOWER) **Goodyear-Nelson Mill No. 2 a few days before the fire. The mill was very active, often ran two shifts.**

Log-feed end of headrig at Goodyear-Nelson Mill No. 2 in June 1946. (TOP) Band saw is stationary while dolly, on rails, with riding operator, moves a log through the cutting cycle. Stationary operator at left controls the ratchet that, when ready, drops next log onto dolly. (LOWER) Same headrig photographed from opposite end. Cut slabs drop onto receiving chain conveyor in foreground, are moved to next mill function.

(TOP) **Vic Nelson (left) and Frank Goodyear, proprietors, had offices in Mill No. 1 which was just outside the Sedro-Woolley city limits as evidenced by the non-dial rural telephone of the Skagit Valley Telephone and Telegraph Company. Telephones inside the city were dial system from West Coast Telephone Company.** (LOWER) **George Lutterloh was office manager, bookkeeper and timber scaler. He is using a hand-cranked full-keyboard Burroughs adding machine. On the desk is a Marchant mechanical calculator. Secretary is unidentified. Wall calendars indicate the day was June 23, 1946, a Wednesday.**

Truck pool at Goodyear-Nelson Mill No. 1. The motor pool served both mills. (LEFT TO RIGHT) **Flat bed, three loggers, flat bed, two dump trucks.**

Some unique items noted in this photograph which was made on March 17, 1947. Murdock St. Sedro-Woolley, Washington. This logging truck, with steerable trailer, was hauling these poles to Puget Sound Light and Power Company. (IN REAR, LEFT) Nelson Chevrolet Garage on corner of Murdock and Woodworth Sts. Tall white building is City Hall, with fire siren on roof. (RIGHT SIDE OF MURDOCK STREET) Skagit Valley Telephone Company next to Burglund Motors (Ford garage). Tall building beyond with peak roof is Eagles Lodge which burned down a few years later. Unseen building with very tall flag pole is American Legion hall. Note open-globe street lights.

114

(TOP) **Kiln dried lumber going through final cutoff which provided boards trimmed to a customer's specifications –"PET" – Precision End Trim).** (LOWER) **The scrap "mill ends" move up conveyor to hopper for storage until a load is dropped into a truck for delivery as fuel to a firewood dealer. The lumber carrier, center, takes a load to the storage yard. Scene is at Goodyear-Nelson Mill No. 3.**

Hand-loading kiln-dried lumber into freight cars at the Port Angeles mill. An office man measures every load carefully, for billing purposes, to determine the number of board feet being shipped. Cars could never be truly fully loaded as there had to be a little "squirm" room for the men working inside the cars.

Circular saw headrig with riding platform at Goodyear-Nelson Hardwood Lumber Company Mill No. 3 at Port Angeles. A winch, seen under the dolly, moves the platform (carriage) and operator back and forth along the short track. Modern safety requirements call for a security rail or cage to enclose the operator and a place to sit. Hard hats were not required and generally unheard of in this period. Man at left guides slabs cut from log on to chain conveyor to next operation. Sawdust often accumulated on the floor until it had to be dug out to continue the work. Lighting in mills was originally from many small windows. When electricity became available, lines were run throughout the mills with light coming from 300 watt bulbs. In picture, note two of the three bulbs are burned out. Burnouts were frequent due to vibrations.

Planers at Goodyear-Nelson
Hardwood Lumber Company mills.
(TOP) **Old planer Mill No. 2.**
(CENTER) **New planers at Mill No. 1.**
(LOWER) **New Planer at Port
Angeles mill No. 3.**

About once every year, usually in the fall, the huge pile of accumulated conky or otherwise defective logs, as well as trash, would be consumed in a bon fire. Such cleanup fires often burned, unattended, for days. No body seemed to mind.

Central Point
Lumber Company

None of the locals seem to remember when this plant was built but, from the nature of the basic construction, it seems to compare with other mills that went up in the late 1920s. Of course many renovations and additions have occurred to accommodate updated methods. For one, there is closed circuit television so the headrig operator can determine if there is a pileup in the conveyor that carries away his output.

The mill is on Highway 99, locally known as Front Street or North Pacific Highway. The present firm, The Central Point Lumber Company, is one of several small mills owned by WTD Industries of Portland, Oregon. WTD is one in a string of operators over the years.

Probably the most notable historical happening to the property occurred on September 25, 1970. On that day, a fire of unknown origin got a start in the cold deck. The firm was then known as Cheney Forest Products. At that time, the log storage area was in Fire District No. 3's (rural) area but trucks from Central Point Rural Fire Protection District also responded as the mill, right

Central Point Lumber Company is within the lightly outlined area in the aerial photograph. It is bounded on the east by Highway 99, The curved city street immediately to the west is Glenn Way. The camera is pointed approximately northeast.

next to the cold deck, was on city property. The firemen hooked hoses to nearby hydrants and started blasting the fire from nozzles on 2½-inch lines. But by now, the fire was too well established and was glowing brilliantly red deep within the deck.

Regrettably, the firemen's work was not enough so a procedural call was made to the Oregon State Forestry Department which in turn placed an emergency request on the U.S. Forest Service for assistance. The Forest Service responded by ordering a fire fighting tanker airplane to take to the air.

The flying distance from the airport to the log deck fire was only about 2 air miles. In the meantime, the Oregon State Police and the Central Point Police Department closed the highway as the stubbornly burning logs were just over the curb from the pavement and hoses had been laid across the lines of traffic. Although the low-level

Central Point Fire Chars Logs, Houses: Loss Undetermined

CENTRAL POINT — Fire Friday destroyed several million board feet of logs at Cheney Forest Products, burned at least two residences, a number of outbuildings and many acres of grass here.

About 80 firefighters representing virtually every fire district and department in southern Jackson County aided in fighting the fires.

While there were three separate fires, the "alarm rang continually," a fireman for the Central Point Rural Fire Department said Saturday. The first alarm was at 1:15 p.m. Friday.

However, the log deck will be difficult to extinguish and will be allowed to burn, under control, until it can finally be put out.

As alarms started coming in Friday at the Central Point fire departments, district boundaries were forgotten. The Cheney fire, which started in the rural department's area, burned across a section of the city and back into the rural department's district.

(TOP) **Forest Service air-tanker makes run on the fire getting ready to drop load of retardant.** (LOWER - LEFT) **Fierce blast of water from Fire Department proved ineffective in controlling the blaze therefore Forest Service was called.** (RIGHT) **Fire burned into the night.**

—Archives, Central Point Fire Dept.

Photograph (TOP) made about 8 p.m. after fire had burned about seven hours. (CENTER) Two days later, little but charred logs remained.
—Archives, Central Point Fire Dept.

bombing of the fire with retardant from the aircraft was expected to be accurately placed, as a precaution all the business buildings along Front Street were evacuated for a distance of about one-half mile.

Within minutes, the airplane came in from the north, dropped to minimum altitude – about 75 feet – and sprayed the burning logs. The tanker pulled up in a climbing turn to prepare for another attack. It seemed only seconds before the plane again buzzed the fire with the second half of its load.

Although the "bombardment" seemed to slow the fire, the airplane returned to the Medford Fire Center for more retardant. During the afternoon, the Forest Service dropped 12,000 gallons of retardant over the blazing logs. All of the log decks were not involved.

As there seemed to be little potential risk of damage to the businesses or to passing vehicles that used the street, Front Street was re-opened. It took several days for the fire to burn itself out.

About 25 additional fires were started, down wind, by falling embers. These fires ranged from small grass fires to structure damage. One of the fire trucks from Central Point and another from State Forestry answered alarms telephoned to the fire house as well as set up a patrol of the streets

The Lonesome Fire Plug

When the City of Central Point put down water lines it included fire hydrants on not-yet-built but on dedicated rights-of-way for streets. The property shown was in the log yard of the lumber company. When the fire started, needed or not the hydrant could not be used because logs were piled around it.

(TOP TO BOTTOM) LaTourneau front-end log loader has just picked up logs in log yard then heads for the feed dock at the mill.

looking for these hot spots. When one was found, these fires were put out then the trucks moved on looking for more trouble.

–0–

In the winter and spring of 1996, Central Point Lumber Company employed 75 to 80 people who worked one shift a day but the work was interrupted from time-to-time because of a shortage of logs. The logs come from private and public stands of timber.

The mill cuts Douglas fir, hemlock, ponderosa pine and lodgepole pine. Its major product is 2 x 4s much of it with precision end trim ("PET") in lengths between eight and ten feet. About eighty percent of its product goes to building contractors with the remaining twenty percent sold to home center markets. The firm ships much of its lumber by rail, but there is a significant portion leaving by truck.

In years past, scrap wood from the several mill processes went up in smoke in a wigwam burner, but now the scraps from Central Point Lumber Company are fed into a chipper at the mill. Chip trucks are loaded under giant drive-through hoppers then the 18-wheel trucks haul their loads mostly to paper-making plants. ◇

123

Portable crane picks up logs one-at-a-time from ready-pile at the feed dock – places each log on conveyor on platform leading to de-barker.

124

(TOP TO BOTTOM) **Logs move on conveyor into moving de-barker line (de-barker not shown) then are carried on next conveyor ...**

...into boxed chain conveyor (the "chute") headed for cut-off saws. Operator will cut logs into pre-determined lengths and trims uneven ends from glass-enclosed operator console. (LOWER) the two cut-off saws and the hour-glass rollers were stopped for picture. These are 46-inch saws with replaceable teeth. The chain-driven hour-glass rollers will advance a log into the saws at operator's discretion.

First saw trims rough end, the end drops through scrap trap.

Log advances on hour-glass rollers, stops, saw makes the cut then log moves on chain conveyor to claw loader (seen on next page) which lifts log to chain conveyor for transport to headrig.

Saws continue to cut-to-length and make final end-trim if needed.

(TOP TO BOTTOM) **Deck loader works like a pitch-fork lifting each log to the platform which conveys the logs into the headrig chain conveyor.** (CENTER) **Logs on the conveyor have been cut in either 8, 9, or 10 foot lengths.** (LOWER) **The headrig. Band saw slices each log into slabs that drop** (RIGHT) **onto the next chain conveyor to first green-chain inspection and sorting worker ...**

(TOP) **First green chain inspector receives boards directly from headrig. Lumber that is good, moves to left over bridge and continues to next inspector.**

(LOWER) **Lumber that is obvious scrap is dumped into boxed chute by operating foot-pedal which raises bridge allowing the scrap to drop, the scrap headed for chipper.**

(TOP) Second green chain inspector hand-removes unacceptable lumber, allows good boards to pass to remilling where lumber is further cut. (CENTER) Steel kiln carts on which semi-finished lumber is hand-stacked, then awaits fork-lift truck ride to kiln. (LOWER) The kiln, where lumber going in may have as much s 48% moisture content – comes out about 40 - 50 hours later at about 19% moisture.

(TOP) Fork lift has removed load of lumber from kiln, sets
it down then (LOWER) picks off one-quarter of the stack
which is then shuttled to the planner for finishing.

Kiln-dried "PET" (Precision End Trimmed) lumber (TOP) banded and stacked awaiting shipment. (CENTER) Non-precision lumber ready for shipping.

(LOWER) This finished 2 x 4 "PET" lumber with manufacturer's logo imprinted on the end of every piece, is on fork-life headed for loading ramp on the railroad siding next to the building.

The failure or success of a business often depends on the availability of parts. In this warehouse, in addition to the large variety of sprockets, are coils of new conveyor chain (lower), as well as motors, switches – all of the "nuts and bolts" needed to keep the shop operating.

The Scraps Become Chips

(LEFT) **Scrap lumber is seen in a boxed conveyor moving into the chipper (beyond wall). The inspector wears safety rope to keep him out of the chipper should he lose his footing while straightening jams in the flow.** (BELOW) **Chips are ideally cut to $5/8$ inch wide by $3/4$ inch long.**

(LOWER) **Drive-through under overhead hoppers services chip trucks which sometimes stand in line waiting to pickup loads.**

The Saw Shop

(ABOVE) In the "good-old-days" of logging, sharpening saws was by hand-filing one tooth at a time. Now, routine sharpening is unattended and automatic. (RIGHT) Saw shop's head filer carries sharpened band saw from shop to the mill.

(ABOVE) On occasion, teeth are knocked out of a saw. When this happens, these small pieces of steel can fly through a mill like shell fragments – can be deadly. The mill's head filer has welded missing teeth at arrows. (LEFT) Specialist straightens saw with a ball pein hammer and steel block.

Oregon Myrtlewood

There are lumber mills and there are *lumber mills*. There are all sizes of lumber mills! Those mills most readily known harvest the big trees that become telephone poles; become building materials for tall and short buildings; mills that peal thin sheets from big logs – veneer – that is made into plywood; mills that cut cedar into bolts from which hand-split roofing shakes are made; the stud mills that produce a seemingly never ending stream of 2 x 4s, 4 x 4s and other basic building blocks of the construction industry.

Then there is The House of Myrtlewood and its specialty mill in Coos Bay, Oregon.

The House of Myrtlewood cuts a unique specie popularly known at "myrtlewood." This wood, (*Umbellularia californica*), is ideal for machine turning. It is outstanding for making furniture especially for table tops, gun stocks, trays, dishes, bowls. Its sur-

A tree that looks at God all day,
And lifts her leafy arms to pray;

...

Poems are made by fools like me,
But only God can make a tree
—Joyce Kilmer

Myrtle Preserves

The myrtle preserves are a series of small tracts along highways on which great numbers of the beautiful, symmetrically shaped globular, evergreen trees are growing. Several of these preserves are in Douglas and Coos Counties along highways 38 and 42 and are easily seen by passing travelers. These trees grow extensively along streams and in the low portions of adjoining hillsides. The Oregon State Parks and Recreation Department maintains these groves.

(LEFT) **Elliptical, lace-shaped leaves and fruit of the myrtle tree.** (RIGHT) **Cold deck at log yard of House of Myrtlewood.**

face polishes to mirrorlike brilliance like no other wood.

This custom mill manufactures "all of the above" in a shop that has been in operation at the same location since 1929. Visitors can tour the shop and watch the various manufacturing processes or they can stop at video locations to see and hear how the wood is worked into goods.

Myrtlewood is a moderately hard wood and quite heavy. Its pores are barely visible with a magnifying glass. This wood is classed as "diffuse porous" as its pores are very uniform. The wood's rays are very narrow but on close scrutiny are visible to the naked eye. The growth rings are separated by dark tissue. The sapwood is cream in color with variations to brown tones. The wide color range in heart-wood runs from tan through reddish-brown to brown.

These trees are medium to large size and grow in Oregon and in California in clearly defined areas. The bark is dark, reddish brown, quite thin with flat scales. This is an evergreen tree with elliptical or lace-shaped leaves two to five inches in length. The tree is wedge-shaped at the base and has a short trunk. The leaves are dark green

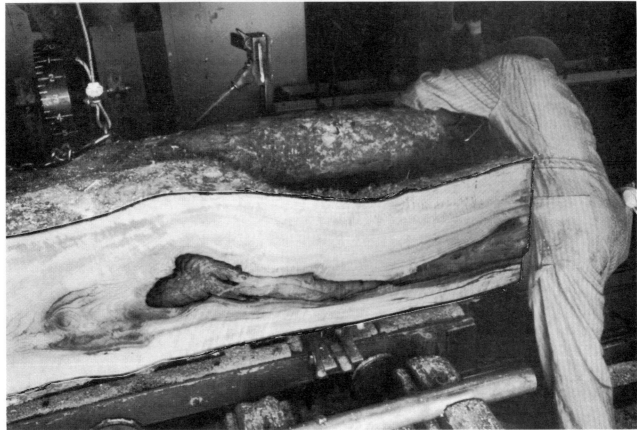

(TOP) **Headrig at the myrtlewood mill with 46-inch diameter circular saw fitted with replace-able teeth. The largest log through this headrig was a 54-inch where a chainsaw was used to complete the cut.** (LOWER) **Chief sawyer takes measurement between cuts.**

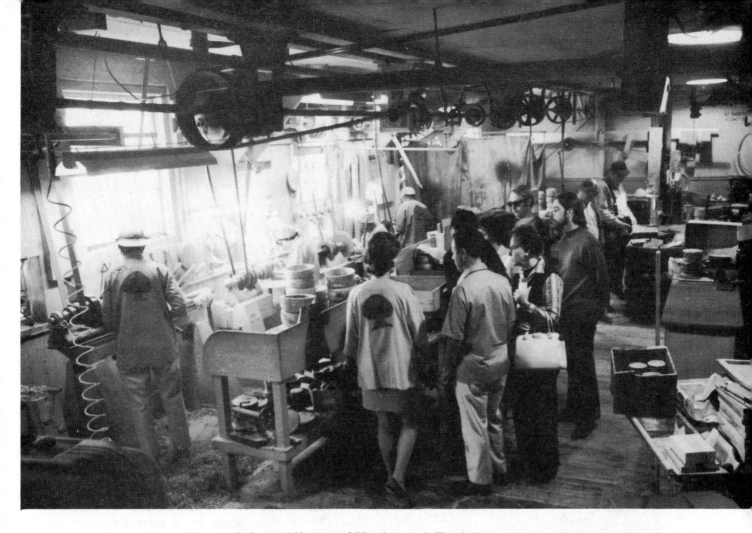

(TOP) **Tour group visits workshop at House of Myrtlewood. The lathes are operated from a "line-shaft" mounted just under the ceiling that works from a single 5 hp motor. Each machine is powered from the shaft with a leather belt. The line-shaft shown was installed in 1929 and has seen continuous use for about 75 years. It is one of the few full-time operating line-shafts in commercial use in Oregon.**

(LEFT) **This mill operates up to nine lathes at a time. Handles above each operator's head is "on/off"** lever to start and stop each lathe. (RIGHT) Scheme illustrating how a line-shaft works for a metal shop. House of Myrtlewood uses multiple wood-turning lathes instead of grinder, shaper and milling machine.

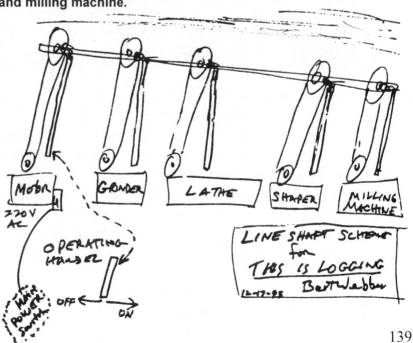

MOTOR   GRINDER   LATHE   SHAPER   MILLING MACHINE

220V AC

OPERATING HANDLE

MAIN POWER SWITCH   OFF ← → ON

LINE SHAFT SCHEME for THIS IS LOGGING

Bert Webber

12-17-95

139

Ed Hodge worked at the mill and retired after fifty years there. He started as a sander, worked his way through the shop eventually becoming the shop foreman.

He pointed out that the circular saw (headrig) is the first saw a log is processed through then small band saws do subsequent cutting.

He related that myrtlewood, being quite hard, must be air-dried in a drying room for about three weeks before use. To dry it in a kiln will harden it to such a point that tooling it on a lathe becomes nearly impossible.

Hodge said that one of the largest items ever made was a 30-inch salad bowl.

and shiny on top and dull on the bottom. When the tree blooms, is produces a yellowish green blossom in clusters, each flower about 3/16 inch long. The fruits (not to be confused with acorns) are rounded and are about one inch diameter. They are green to purplish color. The U. S. Department of Agriculture points out that this tree is neither a laurel nor myrtle but belongs in a family that includes the eastern sassafras.

Woods experts claim the best single word to describe myrtlewood is "beautiful." Leaves of the *Umbellularia californica** are very aromatic when crushed. Some folks use the leaves of the tree for spices in cooking. <>

---

* Also known as Oregon-myrtle, California-laurel, mountain laurel, spice-tree. Refer to bibliography for *Trees*.

Glimpses

Short Takes on Some Other Timber Operations

CABAX Mill in Kerby, Oregon (LEFT) was a general mill working full shifts in the mid 1970's. After the mill ceased to operate, most of the buildings were removed. (RIGHT) One would hardly expect that a mill had been on the grounds except for the wonderment of why is there a square pond there? Mother Earth's ponds do not come with square corners. Today's pond with the square corners was once for logs. (UPPER-RIGHT) Trucker dumps logs directly into the pond.

Medite Corporation is the manufacturer of the world's leading medium density fiberboard made in the plant in Medford (see picture on page 78). The Medite mill (LEFT) in Rogue River, alongside Interstate-5, produces green veneer, operates kilns, and manufacturers chips from logs designated for this product.

141

Timber Products Company (TOP), has mills, sales and distribution offices in Michigan, Oregon, California and British Columbia. The Medford, Oregon location operates a mill that manufactures particleboard, hardwood faced plywood and wood chips. (RIGHT) Dumping a load of sawdust at Medford.

There are three saw mills, one shown (BELOW) and a large number of finishing mills in White City, about ten miles northeast of Medford.

KOGAP Manufacturing Company has a history in the Rogue River Valley dating from about 1940. In the early days, it produced general lumber then shifted to veneer and plywood. Although KOGAP (*Keep Oregon Green and Productive*) has been maintaining tree farms for several decades, the plant closed due to timber shortage. Much of the property, near the south city limits of Medford, on Highway 99, has been converted into a golf course.

There are thousands of acres of Tree Farms throughout the Pacific Northwest. This is one of several belonging to KOGAP.

Tiller Mill and Lumber Company operated its saw mill on the Umpqua - Rogue Divide at a site named Shangri-La after the Shangri-La in James Hilton's book *Lost Horizon*. When close-by timber ran out and it was uneconomical to truck logs to the saw mill, the mill was closed and all of the buildings were removed. Today in looking at the site (RIGHT), one cannot recognize any signs of a mill ever being there.

144

Wood's Accidents

At East Side Logging Company, the railroad box car that had been converted into a crew car had special racks for the saws. Only the men who used a saw were permitted to remove it from the rack and when the train was moving, no saw was touched. The keen sharpness of the teeth could take a man's leg off.

These were the days before hard hats and before the idea of putting steel caps on the tops of the toes of shoes. The issue of safety in the woods was a prime concern for camp superintendents and they harped about safety all the time. There were safety-oriented posters and signs in the cook house. The men knew their lives were at risk if there was a careless worker in their midst and in general, carelessness was not a part of the scene.

Nevertheless, accidents did occur in the forest and if they were serious, they were frequently fatal if for no other reason than lack of medically trained people within reach. These were days of limited communications. There were no "life-line" helicopter rescue services, no land ambulances stationed in the woods or roads over which ambulances could easily pass. Cellular telephones were not even a dream for the future. There was often no way to request help and the men in the camps were not medics.

Broken arms and broken legs seemed to be the most hazardous risk. If a man broke his arm, he was at least mobile and could walk out. A broken leg was another matter. Head injuries were always serious and there were dislocated shoulders. It was rare that a man was killed outright by a falling tree, but this kind of accident was always a risk and such deaths did occur. When bucking a large log into lengths for hauling it from the woods, there was risk of the log rolling and crushing the worker when the final cut was made.

Flying broken cables from various lines in the woods were dangerous. In one case, a logger lost his head by a flying cable. There was no accident insurance and no medical benefits. For his work, in the 1920's, a laborer in the woods may have been paid a dollar or so a day along with room and board.

Injured Logger Safe in Grangeville Hospital After Grueling Trip

The author witnessed an ambulance call for a logger with a broken leg and crushed shoulder that occurred deep in the forest in 1958 along the Salmon River in Idaho near Riggins. The accident happened about 8 on a summer morning. While fellow workers hustled the injured man out of the woods to the cookhouse, a messenger had been dispatched to Riggins on a horse where a long-distance telephone call was made to Grangeville for an ambulance. The ambulance drove to Riggins then took the unimproved road into woods.

The narrow, 2-lane highway No. 95, between Grangeville and Riggins was like riding a corkscrew down the side of White Bird Hill from 4,500 feet elevation, at speeds of not more than about 25 miles an hour, to Riggins at 1,800 feet elevation. On that old road, it was about 50 miles but the trip took nearly three hours.

The ambulance, with its patient, finally emerged from the forest about noon then the attendant and the driver stopped at a Riggins cafe for lunch. The injured man was conscious and was fed some soup after which the trip was resumed. The drive up White Bird Hill was much slower than the trip down. The injured logger finally arrived at the Grangeville hospital late that afternoon. —The author worked for Remington Rand and was in Riggins selling typewriters and adding machines but was also a free-lance reporter for the Lewiston *Tribune*.

The men knew they had to be careful but in spite of the warnings, posters on the cookhouse walls, and individual care in the woods, there were accidents and occasional deaths. When death occurred, that part of the operation had to be closed until the County Coroner could be summoned to the site, a corner's jury enpanaled, and an official hearing conducted. ◇

HIP BONE IS FRACTURED BY FALLING TREE

NANNING CREEK WOODSMAN NARROWLY ESCAPES DEATH

Fred Woodcock, a woodsman employed by the Pacific Lumber company at their Nanning creek camp, near Scotia, was seriously injured yesterday afternoon, when the top of a small tree fell and struck him, breaking his hip and severely bruising his body. He was brought to the Union Labor hospital in Eureka yesterday afternoon for medical treatment.

Woodcock said last night that he was unaware of any danger and the first he knew of it was when he was struck. Unless complications set in it is expected he will recover rapidly.

—The Humboldt Times Aug. 15, 1913. p.1

HEAD TAKEN OFF IN TWO CUTS BY FLYING CABLE

(By Special Correspondence)

WOODSMAN MUTILATED BY ONE CABLE AND DECAPITATED BY SECOND PIECE

FORTUNA, Aug. 3.—That the death of Alfred Fowler, the woodsman killed in the Metropolitan woods Saturday was caused by accident caused by the breaking of a cable attached to a Tommy Moore and that no blame was attached to anyone, was the verdict of the jury empannelled by Deputy-Coroner E. J. Hunter. Those comprising the jury were Lou Barry, C. Black, D. S. Chamley, H. T. Martin, V. A. Harkins and P. W. Hunter.

At the time of the accident Fowler was seated on a log waiting for the cable to bring the logs along. The line broke and struck him on the head just below the eyes. The upper portion of the head was sliced off and a minute later another portion of the line struck him and cut the remaining portion of the head off at the neck.

Fowler was about 40 years of age and leaves a widow and five children in Finland, of which country he was a native. A brother-in-law, Erick North, resides at Stitz creek. Until he is heard from no arrangements will be made for the funeral, though it is probable it will be held Monday.

—The Humboldt Times Aug. 7, 1913. p.1

The Japanese Failure

"…you are to drop bombs and start forest fires to burn down the timber industry of the Pacific Northwest"

— orders to Japanese pilot Nubuo Fujita on his leaving Japan by submarine to bomb a forest in Oregon in World War-II.

The Japanese sought to burn down all the forests of the Pacific Northwest during a six-month blitz late in World War-II. Due to the usual wet weather during the offensive, the trees were wet and would not burn. The first attack came on September 9, 1942 when a piloted aircraft with two Japanese aboard, dropped bombs in the Siskiyou National Forest of Oregon and started a small fire. Having proved that it was possible to launch a bombing airplane from a *submarine*, the Japanese planned to repeat the manned-bomber attacks but could not when the Imperial Navy pulled out of the project because of severe losses at the Battle of Guadalcanal. The Imperial Army, however, set out to develop trans-continental free-floating balloons that would cross the Pacific Ocean in just a few days in the jet stream.

These balloons, 70 feet high and full of deadly hydrogen, each carried five bombs, four of which were incendiaries. The fifth bomb was a high-explosive anti-personnel bomb. Documents reveal that a minimum of 30, 000 bombs were launched against the western United States in the period from November 1944 until mid-March 1945. One of these bombs killed kids and an adult in Oregon who were on a picnic-hike in the Fremont National Forest near the village of Bly. These were the only civilian deaths in the continental United States caused by direct enemy action during the war.

To the present time, 369 parts of bombs and balloon apparatus have been discovered in 28 states, Canadian provinces and in Mexico. The largest numbers of "balloon incidents" are 45 in Oregon; 32 in Montana; 28 in Washington, 25 in California, 12 in Idaho. The farthest east where parts have been recovered is Detroit, Michigan. The most recent discovery was in Oregon in summer 1992.

The project was very costly to the United States and Canada because of the number of men and women that were busy keeping track of and retrieving balloons and their parts (some landed intact) as discoveries were reported. A number were shot down by fighter planes. But because the forests were wet, the Japanese balloon bombing offensive to burn down the forests was a grand failure.

(RIGHT) **Mt. McLoughlin, 9, 495 feet elevation attracted a number of Japanese bombing balloons. The forests still stand because they were wet and would not burn.**

The book that is the international authority on the balloon bombs and other Japanese attacks on the Unites States mainland, which the majority of people still know little or nothing about, is *Silent Siege-III*. Refer to bibliography.

Re-seeded forest areas earlier clear-cut. Trees, like many other agricultural products, are a renewable crop.

The Pacific Northwest Will
Never Grow Out of Trees!

Photograph made during the Webbers' visit to the log yard at Rogge Forest Products Co., Bandon, Oregon.

About the Authors

Bert and Margie Webber have been enjoying and writing about the Pacific Northwest for many years and have written a number of books together. Their work is non-fiction and has been found enjoyable by history buffs, teachers, and for use by librarians and scholars. The books are written with authority but in easy-to-read language. All have bibliographies and indexes. This is their second book about logging. During the past many years, they have never lived or worked very far from a saw mill or other timber operation.

Bert Webber is a Research Photojournalist. He holds a degree in journalism from Whitworth College and earned the Master of Library Science degree with studies at Portland State University and the University of Portland.

He spent many years as a Commercial Photographer and photo-lab specialist, having earlier served as an Official Signal Corps Photographer in the World War II. He has written many dozens of newspaper and magazine feature articles and several dozen non-fiction books. His subjects are mostly about the Pacific Northwest – especially Oregon – and unique matters of World War II in the Pacific.

Margie Webber, is a retired Registered Professional Nurse. She holds a degree in Nursing from the University of Washington. She enjoys geography and meteorology and often serves as a field photographer. Her main work with this book is as a field assistant and editor.

Bert and Margie Webber relax with music of which they enjoy a wide variety but especially pipe organ. Bert plays Euphonium in the Southern Oregon Symphonic Band and serves on the band's Board of Control. Margie plays piano and organ and is a member of her church's English Hand-Bell Choir.

The Webbers have four children and eight grandchildren. They live in Oregon's Rogue River Valley in the once-a-village, now-the-City of Central Point where their home is only ten blocks from a saw mill. <>

Bibliography

—Books

Andrews, Ralph W., *"This Was Logging."* Superior. 1954.

Armstrong, Chester H. Ed. *Oregon State Parks History.* State Printer (Salem). 1965.

Hauff, Steve and Jim Gertz. *The Willamette Locomotive.* Binford & Mort. 1977.

LaLande, Jeffery M. *Medford Corporation: A History of an Oregon Logging and Lumber Company.* (private print). 1979.

McCullough, Walter F. *Woods Words; A Comprehensive Dictionary of Loggers Terms.* Oregon Hist. Soc. 1953.

Merriam, Lawrence C. Jr. *Oregon's Highway Park System 1921-1989...; Includes. Historical Overview and Park Directory.* Oreg. State Parks and Rec. Dept. 1992.

Pierre, Joseph H. *When Timber Stood Tall.* Superior. 1979.

Robertson, Donald B. *Encyclopedia of Western Railroad History.* Caxton. 1995.

Trees; The Yearbook of Agriculture – 1949. U. S. Dept. of Agriculture. U. S. Gov. Print office. 1949.

Webber, Bert. *Battleship Oregon; Bulldog of the Navy.* Webb Research Group. 1994.

_____. *Silent Siege-III. Japanese Attacks on North America in World War II— Ships Sunk, Air Raids, Bombs Dropped, Civilians Killed.* Webb Research Group. 1992

_____. *Swivel-Chair Logger.* YeGalleon. 1976.

—Newspaper and Magazine articles

"Aircraft Ditched As Fuel Runs Out" in *Mail Tribune.* July 28, 1970.

"Blaze Near Talent Chars 130 Acres" in *Mail Tribune.* July 25, 1977. P-l.

"Central Point Fire Chars Logs, Houses: Loss Undetermined" in Medford *Mail Tribune.* Sept. 27, 1970. P.1.

Fattig, Paul. "Snow Logging Kinder to Forest; White Stuff Cushions Fragile Land" in *Mail Tribune.* Jan. 25, 1996. P. 1.

Fattig, Paul. "BLM Will Restrict Sale Access" in *Mail Tribune.* Jan. 25, 1996. P.2A

_____. "BLM Won't Block Off Hoxie Sale; Croman Corp. Begins Timber Harvest today in *Mail Tribune.* Jan. 24, 1996. P. 1.

Fattig, Paul. "Protest Planned For Logging Near Howard Prairie Lake" in *Mail Tribune.* Jan. 27, 1996.

Force, David. "Medford's Fire Retardant Tanker Base" in *Mail Tribune.* Aug. 12, 1973. P.

Gregory, Gordon. "Protesters Delay Hoxie Loggers" in *Mail Tribune.* Feb. 3, 1996. P.1A.

Hallmark, Allen. "Air Tanker Flying: It's A Lonely, Dangerous Business" in *Mail Tribune.* July 25, 1976. P-1.

Manny, Bill. "Fireman Recalls Work Days on 'Sidewinder" in *Mail Tribune.* June 30, 1986.

"270-Acre Logtown Area Fire Reported Controlled in *Mail Tribune.* Aug. 30, 1972

Webber, Bert. "Medford Firm Converts old DC6Bs into 'Bombers'" in *Oregon Journal.* Aug. 28, 1972. P-12

_____. "Medford 'Battle Squadron' Carries War to Forest Enemy" in *Oregon Journal.* Aug. 18, 1970.

Index

Page numbers for illustrations are shown in **bold italic** type

W

Webber, Bert, *33*, 139, *150*
Webber, Ebbert T., 105
Webber, Margie, *150*
Weyerhaeuser Timber Co., 7
White City (town), Ore., mills, *142*
wigwam burner, 9, *17*
Willamette Iron and Steel Works, 10,
    53-54, 55
Willamette Shay, 9, 13, 14, *15, 53, 54,*
    55, 69, *79*, 82, *83, 84, 85,* 86;
    superheated, 86; wrecks, *60, 61*
Willamette Shay No. 4, *83, 84, 85*
wire rope, *63*
"Wobblies," defined 9, 17; *see also*:
    I.W.W.
women in camp. 46, 47, 49, 50
World War-II bombing of forest by
    Japanese, 147
wrecks, *60, 61*, 67, 69, *70*

Y

yarding, 9; 36; defined, 18

—Notes—

—Notes—